SOLAR
ELECTRICITY

UNESCO ENERGY ENGINEERING SERIES
ENERGY ENGINEERING LEARNING PACKAGE

Dr. Boris Berkovski
Chairman of the Editorial Board
Director, Division of Engineering and Technology
UNESCO

Organised by UNESCO, this innovative distance learning package has been established to train engineers to meet the challenges of today and tomorrow in this exciting field of energy engineering. It has been developed by an international team of distinguished academics co-ordinated by Dr Boris Berkovski. This modular course is aimed at those with a particular interest in renewable energy, and will appeal to advanced undergraduate and postgraduate students, as well as practising power engineers in industry. UNESCO gratefully acknowledges the financial contributions and other forms of cooperation received from the International Technological University (ITU), London and Paris, over the period 1991-94 towards several of the texts in this series. Without the consent and goodwill of the ITU Board of Trustees, and especially Dr S.M.A. Shahrestani, one of its founding members, some of the texts could not have been produced.

Solar Electricity
Edited by Tomas Markvart

Magnetohydrodynamic Electrical Power Generation
Hugo K. Messerle

Geothermal Energy
Edited by Mary H. Dickson and Mario Fanelli

Energy Planning and Policy
Maxime Kleinpeter

SOLAR
ELECTRICITY

Edited by
Tomas Markvart

University of Southampton
UK

with

Klaus Bogus *(ESTEC – ESA, The Netherlands)*
Luis Castañer *(UPC, Barcelona, Spain)*
Andres Cuevas *(IES – UPM, Madrid, Spain)*
Philip Davies and **Juan Carlos Miñano** *(IES – UPM, Madrid, Spain)*
Miguel Angel Egido *(IES – UPM, Madrid, Spain)*
Robert Hill *(University of Northumbria, UK)*
Ben Hill *(Hiltec Solar, UK)*
Oliver Paish, Bernard McNelis and **Anthony Derrick** *(IT Power, Eversley, UK)*
Eduardo Lorenzo *(IES – UPM, Madrid, Spain)*
Sean McCarthy and **Martin Hill** *(Hyperion, Cork, Ireland)*
Aleksei Sorokin *(TEAM, Rome, Italy)*
Michael Specht *(ZSW, Stuttgart, Germany)*

JOHN WILEY & SONS
Chichester • New York • Brisbane • Toronto • Singapore

Copyright © 1994 UNESCO

Published in 1994 by John Wiley & Sons Ltd,
                    Baffins Lane, Chichester,
                    West Sussex PO19 1UD, England
                    National Chichester (0243) 779777
                    International (+44) 243 779777

Reprinted February 1995

Other Wiley Editorial Offices

John Wiley & Sons, Inc., 605 Third Avenue,
New York, NY 10158-0012, USA

Jacaranda Wiley Ltd, 33 Park Road, Milton,
Queensland 4064, Australia

John Wiley & Sons (Canada) Ltd, 22 Worcester Road,
Rexdale, Ontario M9W 1L1, Canada

John Wiley & Sons (SEA) Pte Ltd, 37 Jalan Pemimpin #05-04,
Block B, Union Industrial Building, Singapore 2057

British Library Cataloguing in Publication Data

A catalogue record for this book is available from the British Library

ISBN 0 471 94161 1

Typeset in 10/12pt Times by Pure Tech Corporation, Pondicherry, India
Printed and bound in Great Britain by Redwood Books, Trowbridge, Wiltshire

Contents

Foreword

Education applied to a whole complex of interlocking problems is the master key that can open the way to sustainable development.

A major obstacle to the development of renewable energies is the paucity of relevant information available to engineers and technicians (who too often lack the necessary know-how and skills) as well as to decision-makers and users. One of UNESCO's priorities is this domain is therefore to promote training and information aimed at sensitizing specialists and the general public to the possible uses of renewable energy sources, with particular regard to environmental concerns and the requirements of sustainable development.

The present textbook on renewable energies in UNESCO's Energy Engineering Series is designed not only to provide instruction for new generations of engineers but also to foster the kind of environmental management so urgently needed throughout the world. The series as a whole is a co-operative venture involving many institutions working together—on the basis of pre-defined standards—to promote awareness of the environmental, cultural, economic and social dimensions of renewable energy issues as well as knowledge of those technical aspects that form the core of the UNESCO/Learning Package Programme.

The World Solar Summit that met at UNESCO Headquarters in July 1993 made education and training one of its strategic priorities. This Series should make a useful contribution to the World Solar Summit Process, and UNESCO is pleased to be a partner in its development.

Federico Mayor
Director-General
UNESCO

Introduction

This book deals with photovoltaics: the direct conversion of light into electricity. It is based on the Short Course on Solar Energy Conversion and Applications which takes place annually at the University of Southampton but has been expanded to provide a self-contained document suitable for distance learning.

A brief overview of the subject is given in Chapter 1, devoted to the history, present status and future of this increasingly important renewable energy source.

In Chapter 2 we show how to determine the supply of solar energy for a given location. The geographical effects are analysed, and it is shown how the available solar radiation data can be used to determine the useful energy supply.

Chapter 3 explains how solar cells work and how they are manufactured. The principles of photovoltaic energy conversion are here discussed alongside the practical solar cell operation. The manufacture of the solar cells which are most commonly used at present and those entering the market—crystalline silicon and thin films—is described in some detail.

The photovoltaic power system is discussed in Chapter 4 which looks at the system structure and subsystems. This unit also includes the system sizing and design.

Applications of solar electricity are reviewed in Chapter 5, together with the economics of PV installations. Attention is focused on the various applications of photovoltaics which are relevant at present and also on those that are likely to to be of increasing importance in the near future.

The environmental and social impacts of energy production are considered in Chapter 6 where the methodology for environmental assessment of energy sources is described and applied to photovoltaic power generation.

Important specialised topics are treated in Chapter 7 which includes the discussion of large photovoltaic systems, photovoltaics under concentrated sunlight, and the hydrogen economy.

Acknowledgements

The preparation of this text has been a team effort. Many experts in various aspects of photovoltaics have shared their knowledge with me and, with their help, I have moulded their contributions into the format before you. All these contributions have been of a very high standard and I accept full responsibility for any inadvertent errors or ommissions in the printed text.

Professor Arthur Willoughby, Mr Fred Treble, Professor Peter Landberg and Mr John Bonda who give lectures at the Southampton course have conveyed much information to me through their lecture notes, class work and discussions, and their input is reflected in this module.

Professor Israel Dostrovsky has, through the kind gift of his book, given me much insight into the global aspects of energy production. Professor Ray Arnold has patiently explained to me the structure of electricity transmission and distribution systems—in particular, the meaning of the term 'grid'. I am also indebted to Mr Paul Maycock, the Directorate General for Energy of the Commission of the European Communities, and all contributors of the data and photographs that are reproduced in the text.

I would like to express particular thanks to Professors Antonio Luque, Gabriel Sala, Gerardo Araujo, Eduardo Lorenzo, and all their colleques and students, for many stimulating discussions and hospitality during my stay at the Instituto de Energia Solar in Madrid where most of this text was prepared. The help of Mr Ricardo Castrillo with tracing biblography and of Mrs Montserrat Rodrigo with the graphics is also much appreciated.

Tomas Markvart
Madrid
June 1992

List of Contributors

Dr Klaus Bogus

ESTEC / ESA
PO Box 299
2200 AG Noordwijk
The Netherlands

Professor Luis Castañer

Departamento de Ingeniería Electrónica
E.T.S.I. Telecommunicación
Universidad Politecnica de Cataluña
Calle Gran Capitán s/n
08034 Barcelona
Spain

Ben Hill

Hiltec Solar Ltd
25 Low Friar Street
Newcastle upon Tyne
NE1 5UE
UK

Professor Robert Hill

Newcastle Photovoltaics Applications Centre
University of Northumbria
Ellison Place
Newcastle upon Tyne NE1 8ST
UK

Professor Eduardo Lorenzo, Dr Andres Cuevas, Dr Miguel Angel Egido
Dr Philip Davies and Dr Juan Carlos Miñano
Instituto de Energia Solar
Universidad Politecnica de Madrid
E.T.S.I. Telecommunicación
Ciudad Universitaria
18040 Madrid
Spain

Dr Sean McCarthy and **Martin Hill**

Hyperion Energy Systems Ltd
Enterprise Centre
North Mall
Cork
Ireland

Bernard McNellis, Anthony Derrick and **Oliver Paish**

IT Power
The Warren
Bramshill Road
Eversley, HANTS RG27 0PR
UK

Ing. Aleksei Sorokin

via del Buon Ricovero, 10
00189 Rome
Italy

Dr Michael Specht

Solar and Hydrogen Energy Research Centre (ZSW)
Allmandring 19
D-7000 Stuttgart 80
Germany

Sources of Illustrations and Tables

1: Electricity from the Sun
Table 1.1 *P. T. Landsberg and F. C. Treble*, Fig. 1.1 *I. Dostrovsky*, Fig. 1.2 *P. D. Maycock.*

2: Solar radiation
Table 2.2 *L. C. Castañer*, Figs. 2.1, 2.6, 2.7 *E. Lorenzo*, Fig. 2.4 *F. Kreith and J. F. Kreider.*

3: Solar Cells
Fig. 3.23 *T. F. Ciszek*, Figs. 3.24–3.28, 3.30, 3.33 *P. Hersh and K. Zweibel*, Fig. 3.32 *R. J. Van Overstraeten and R. P. Mertens*, Figs. 3.34–3.41 *R. Hill.*

4: Photovoltaic Engineering
Table 4.2 *E. Lorenzo*, Table 4.3 *A. Sorokin*, Tables 4.4–4.6 *L. Castañer*, Table 4.7 *M. A. Egido*, Fig. 4.3 *J. C. Miñano*, Figs. 4.5, 4.13, 4.15, 4.16 *E. Lorenzo*, Figs. 4.4, 4.18–4.20, 4.22, 4.23 *L. Castañer*, Fig. 4.17 *A. Sorokin*, Fig. 4.21 *H. L. McComber et al.*, Fig. 4.24 *M. A. Egido and E. Lorenzo.*

5: Applications
Table 5.1 *Directorate General for Energy, Commission of the European Communities*, Tables 5.2–5.5 *IT Power*, Table 5.6 *Hiltec Solar*, Figs. 5.1, 5.2, 5.5–5.10a, 5.11–5.15, 5.16c, 5.24 *IT Power*, Figs. 5.3, 5.4, 5.25–5.28 *Hyperion*, Fig. 5.10b *Midway Labs Inc.*, Figs. 5.16a and b *Hiltec Solar*, Figs. 5.17–5.21 *K. Bogus*, Fig. 5.23a *A. Sorokin*, Fig. 5.23b *P. D. Maycock.*

6: Environmental Impacts of Photovoltaics
Figs. 6.1–6.4 *R. Hill.*

7:   Advanced and Specialised Topics
Tables 7.1–7.4 *A. Sorokin*, Table 7.5 *M. Specht*, Fig. 7.1 *P. D. Maycock*, Fig 7.2 *A. Sorokin*, Figs. 7.3, 7.5–7.7, 7.9, 7.10 *P. A. Davies and J. C. Miñano*, Fig. 7.4 *E. C. Boes and A. Luque*, Fig. 7.8, 7.13 *SEA Corporation*, Fig 7.11 *Midway Labs Inc.*, Fig. 7.12a, b *Alpha Solarco*, Figs. 7.14–7.18 *M. Specht.*

Note on Energy Units

We have endeavoured to use the SI system of units as much as possible. However, in two instances it has been expedient (and certainly preferable from the didactic viewpoint) to employ 'other' energy units, as defined by the SI convention.

Firstly, the usual unit of energy in solid-state physics is the electronvolt (eV). This unit is also in common use when discussing semiconductor solar cells.

Secondly, the unit watthour (Wh) and its derivatives (kWh, etc), which are in common use in power engineering, have been used for solar irradiation. This usage is also recommended within the European Community's Solar Energy R&D programme.

The following conversion factors may be found useful.

Factor	Prefix	Symbol
10^{-6}	micro	μ
10^{-3}	milli	m
10^{-2}	centi	c
10^{3}	kilo	k
10^{6}	mega	M
10^{9}	giga	G
10^{12}	tera	T
10^{18}	exa	E

$1 \text{ W s} = 1 \text{ J}$
$1 \text{ Wh} = 3.6 \text{ kJ}$
$1 \text{ EJ} = 278 \text{ TWh}$
$1 \text{ eV} = 1.602 \times 10^{-19} \text{ J}$

1

Electricity from the Sun

1.1 WHY WE NEED PHOTOVOLTAICS

The Sun has been worshipped as a life-giver to our planet since ancient times. The industrial ages gave us the understanding of sunlight as an energy source. This discovery has never been more important than now as we realise that the exploitation of fossil energy sources may be affecting the planet's ambient.

The energy supply from the Sun is truly enormous: on average, the Earth's surface receives about 1.2×10^{17} W of solar power. This means that in less than one hour enough energy is supplied to the Earth to satisfy the entire energy demand of the human population over the whole year. Indeed, it is the energy of sunlight assimilated by biological organisms over millions of years that has made possible the industrial growth as we know it today. Most of the other renewable means of power generation also depend on the Sun as the primary source: hydroelectric, wind and wave power all have the same origin (Fig. 1.1).

The history takes us back over 150 years when, in 1839, Edmund Becquerel observed that 'electrical currents arose from certain light-induced chemical reactions' (Table 1.1). A similar effect was observed in a solid (selenium) several decades later. A comprehensive understanding of these phenomena, however, had to await the progress of science towards the quantum theory in the early parts of this century. The development of the first solid-state devices in the late forties then paved the way to the announcement of a silicon solar cell with 6% efficiency (Chapin *et al.*, 1954)—the first usable solar cell.

Solar cells did not have to wait long to find application. The year 1958 witnessed the launch of Vanguard 1, the first satellite to use electricity from the sun. The technology has been developing ever since. Much interest in solar electricity appeared particularly in the wake of the oil crisis in the

1

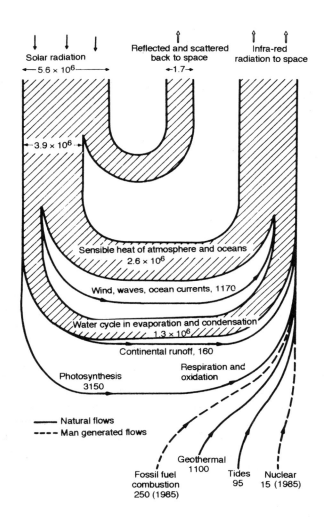

Fig. 1.1 Energy balance of the Earth (EJ/year) (from 1. Dostrovsky, *Energy and the Missing Resource*, Cambridge University Press, Cambridge, 1988, p. 83. Reproduced with permission)

early seventies. Today, the direct conversion of light into electricity, or *photovoltaics*, is becoming accepted as an important form of power generation.

Over 50 megawatts of photovoltaic power modules were produced in 1991 (Fig. 1.2). The production rate has been increasing at almost 20% annually over the last few years, and this trend is likely to continue. As a conservative estimate, the annual module production rate will reach hundreds of megawatts by the end of the century, and tens of gigawatts in the next 40–50 years.

Table 1.1 Some dates of relevance to photovoltaic solar energy conversion†

1839	Becquerel discovers the photogalvanic effect
1876	Adams and Day observe photovoltaic effect in selenium
1900	Planck postulates the quantum nature of light
1930	Quantum theory of solids proposed by Wilson
1940	Mott and Schottky develop the theory of solid-state rectifier (diode)
1949	Bardeen, Brattain and Shockley invent the transistor
1954	Chapin, Fuller and Pearson announce 6% efficient silicon solar cell
1954	Reynolds *et al.* report solar cell based on cadmium sulphide
1958	First use of solar cells on an orbiting satellite Vanguard 1

†After P. T. Landsberg and F. C. Treble, lecture notes at the Southampton Short Course on Solar Energy Conversion and Applications.

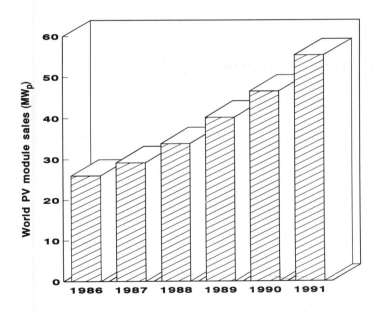

Fig. 1.2 World photovoltaic module sales (in MW$_p$). (Data courtesy of P. D. Maycock, PV News.)

Photovoltaic power generation is reliable, involves no moving parts, and the operation and maintenance costs are very low. The operation of a photovoltaic system is silent, and creates no atmospheric pollution. Photovoltaic systems are modular, and can be quickly installed. Power can be generated where it is required without the need for transmission lines.

There are already a number of terrestrial applications today where photovoltaics provides a viable means of power generation. Many of these installations operate in locations where other means of electricity supply would be out of the question, providing important social benefits to rural communities throughout the world.

Solar electricity suffers one major drawback at present, and that is the high capital cost of photovoltaic installations. Nevertheless, as the production volumes continue to grow and research brings in new developments, the impact of photovoltaics on power generation is set to increase at a rapid rate. The economic comparison with conventional energy sources is certain to receive a further boost as the environmental and social costs of power generation are included in the picture.

Summary

The history, present status and future of photovoltaics have been outlined.

BIBLIOGRAPHY AND REFERENCES

CHAPIN, D. M., FULLER, C.S and PEARSON, G. L., A new p-n junction photocell for converting solar radiation into electrical power, *J. Appl. Phys.* **25**, 1954: 676–677.
DOSTROVSKY, I. *Energy and the Missing Resource*, Cambridge University Press, Cambridge, 1988.

2

Solar Radiation

AIM

The aim of this unit is to determine the solar energy available to photovoltaic systems.

OBJECTIVES

After completing this unit you should be able to:

1. Understand the nature of solar radiation at different locations on the Earth.
2. Describe the apparent daily and yearly motion of the Sun.
3. Show how the amount of solar radiation falling on photovoltaic panels can be determined from the measured data.

NOTATION AND UNITS

Symbol		SI unit	Other unit
B	Daily beam irradiation on a horizontal plane (monthly mean)	J/m^2	Wh/m^2
$B(\beta)$	Daily beam irradiation on a plane at angle β (monthly mean)	J/m^2	Wh/m^2
B_o	Daily extraterrestrial irradiation on horizontal plane	J/m^2	Wh/m^2
D	Daily diffuse irradiation on a horizontal plane (monthly mean)	J/m^2	Wh/m^2

Edited from manuscripts of E. Lorenzo (Solar radiation) and L. Castañer (Photovoltaic engineering).

(*Contd.*)

Symbol		SI unit	Other unit
$D(\beta)$	Daily diffuse irradiation on a plane at angle β (monthly mean)	J/m^2	Wh/m^2
G	Daily global irradiation on a horizontal plane (monthly mean)	J/m^2	Wh/m^2
$G(\beta)$	Daily global irradiation on a plane at angle β (monthly mean)	J/m^2	Wh/m^2
S	Solar constant	W/m^2	
K_T	Clearness index		
$R(\beta)$	Daily irradiation reflected from the ground onto a plane at angle β	J/m^2	Wh/m^2
α	Solar elevation		
β	Panel inclination to the horizontal plane		
δ	Solar declination		
θ_z	Zenith angle		
λ	Wavelength	m	
ϕ	Geographical latitude		
ρ	Reflectivity of the ground		
ψ	Azimuth		
ω	Hour angle		
ω_s	Sunrise hour angle		
	Irradiance	W/m^2	
	Spectral irradiance	W/m$^2\mu$m *or* W/m^{2}eV	

Unit conversion factors

To convert a quantity given in Wh/m^2 to J/m^2, it should be multiplied by 3600.

Values of physical constants

$S = 1367$ W/m^2

2.1 INTRODUCTION

The design of a photovoltaic system relies on a careful assessment of solar radiation at a particular site. Although solar radiation data have been recorded for many locations in the world, they have to be analysed and processed before a sufficiently accurate estimate of the available solar radiation for a photovoltaic system can be made.

This chapter reviews the properties of solar radiation on Earth and outlines the calculation which must be carried out to determine the amount of radiation falling on the PV modules.

First, different components of the radiation will be introduced, and a broad assessment of the global availability of solar radiation will be made.

The astronomical relationship between the Sun and the Earth will then be described. We shall define the relevant quantities which characterise the apparent motion of the Sun.

Finally, we take a closer look at a procedure which makes it possible to determine the amount of radiation available to PV modules at an inclined position from the available solar radiation data on a horizontal surface.

2.2 ENERGY FROM THE SUN

2.2.1 Introduction

To a good approximation, the sun acts as a perfect emitter of radiation (*black body*) at a temperature close to 5800 K. The resulting (average) energy flux incident on a unit area perpendicular to the beam outside the Earth's atmosphere is known as the *solar constant*:

$$S = 1367 \text{ W/m}^2 \tag{2.1}$$

In general, the total power from a radiant source falling on a unit area is called *irradiance*.

When the solar radiation enters the Earth's atmosphere (Fig. 2.1), a part of the incident energy is removed by *scattering* or *absorption* by air

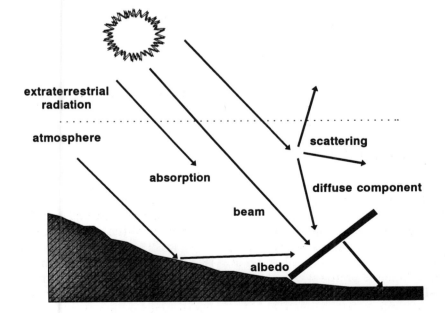

Fig. 2.1 Solar radiation in the atmosphere

molecules, clouds and particulate matter usually referred to as aerosols. The radiation that is not reflected or scattered and reaches the surface directly in line from the solar disc is called *direct* or *beam radiation*. The scattered radiation which reaches the ground is called *diffuse radiation*. Some of the radiation may reach a receiver after reflection from the ground, and is called the *albedo*. The total radiation consisting of these three components is called *global*.

The amount of radiation that reaches the ground is, of course, extremely variable. In addition to the regular daily and yearly variation due to the apparent motion of the Sun, irregular variations are caused by the climatic conditions (cloud cover), as well as by the general composition of the atmosphere. For this reason, the design of a photovoltaic system relies on the input of measured data close to the site of the installation.

A concept which characterises the effect of a clear atmosphere on sunlight is the *air mass* (Fig. 2.2), equal to the relative length of the direct beam path through the atmosphere. On a clear summer day at sea level, the radiation from the sun at zenith corresponds to air mass 1 (abbreviated to AM1); at other times, the air mass is approximately equal to $1/\cos \theta_z$, where θ_z is the zenith angle which will be discussed more fully in the next section.

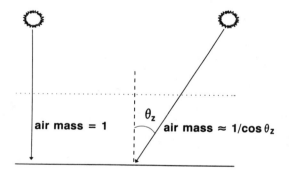

Fig. 2.2

The effect of the atmosphere (as expressed by the air mass) on the solar spectrum is shown in Fig. 2.3. The extraterrestrial spectrum, denoted by AM0, is important for satellite applications of solar cells. AM1.5 is a typical solar spectrum on the Earth's surface on a clear day which, with total irradiance of 1 kW/m^2, is used for the calibration of solar cells and modules. Also shown in Fig. 2.3 are the principal absorption bands of the air molecules.

Although the global irradiance can be as high as 1 kW/m^2, the available irradiance is usually considerably less than this maximum value because of the rotation of the Earth and adverse weather conditions. For illustration,

Fig. 2.3 The solar spectrum

Fig. 2.4 shows the mean annual irradiance in different parts of the world. The mean irradiance is highest near the latitudes of the tropics of Cancer

Fig. 2.4 The mean annual irradiance on a horizontal plane (in W/m^2) across the world (from F. Kreith and J. F. Kreider *Principles of Solar Engineering*, Hemisphere Publishing Corp., New York, 1978, page 17. Reproduced with permission)

and Capricorn, but is lower in equatorial regions on account of the cloud cover. In higher latitudes, of course, solar radiation is weaker because of low solar elevation.

Fig. 2.5 The annual profile of solar radiation as a function of geographical location

Solar irradiance integrated over a period of time is called *solar irradiation*. Of particular significance in the design of photovoltaic systems is the irradiation over one day. The yearly variation of both the global and diffuse daily irradiation on a horizontal plane for three locations ranging from the Sahara desert to northern Europe is shown in Fig. 2.5. Note that the seasonal variation becomes more pronounced with increasing latitude.

2.2.2 Some astronomy

The Earth revolves around the Sun in an elliptical orbit (very close to a circle) with the Sun in one of the foci. The plane of this orbit is called the *ecliptic*. The time taken for the Earth to complete this orbit defines a year. The relative position of the Sun and Earth is conveniently represented by means of the *celestial sphere* around the Earth (Fig. 2.6). The equatorial plane intersects the celestial sphere in the *celestial equator*, and the polar axis in the *celestial poles*. The motion of the Earth round the Sun is then pictured by apparent motion of the Sun in the ecliptic which is tilted at 23.45° to the celestial equator. The angle between the line joining the centres of the Sun and the Earth and the equatorial plane is called the *solar*

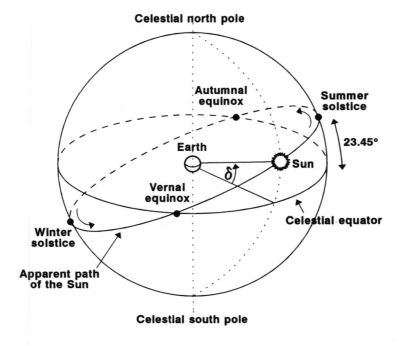

Fig. 2.6 The celestial sphere with the apparent yearly motion of the Sun

declination and denoted by δ. This angle is zero at the *vernal* (20/21 March) and *autumnal* (22/23 September) *equinoxes*. On these days, the sun rises exactly in the east and sets exactly in the west. At the *summer solstice* (21/22 June), δ = 23.45° and at the *winter solstice* (21/23 December), δ = − 23.45°.

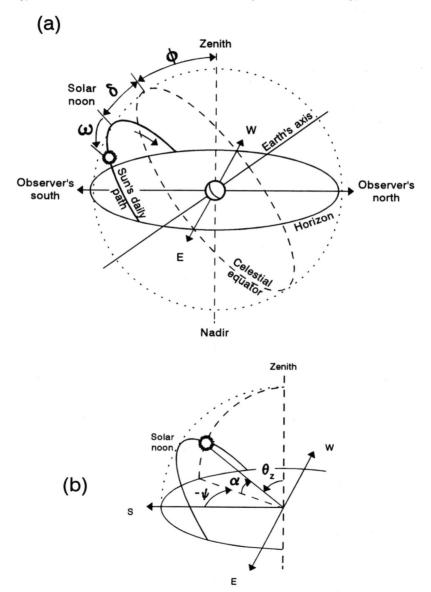

Fig. 2.7 (a) The local zenith–nadir coordinate system showing the apparent daily motion on the Sun. (b) Definition of the azimuth, solar elevation and the zenith angle

The Earth itself rotates, at the rate of one revolution per day, about the *polar axis*. The daily rotation of the Earth is depicted by the rotation of the celestial sphere about the polar axis, and the instantaneous position of the sun is described by the *hour angle* ω: the angle between the meridian passing through the sun and the meridian of the site. The hour angle is zero at solar noon and increases towards the east (Fig. 2.7a). For an observer on the Earth's surface at a location with geographical latitude ϕ, a convenient coordinate system is defined by a vertical line at the site which intersects the celestial sphere in two points, the *zenith* and the *nadir*, and subtends the angle ϕ with the polar axis. The great circle perpendicular to the vertical axis is the *horizon*. The angle between the sun's direction and the horizon is the *elevation* α whose complement to 90° is the *zenith angle* θ_z. The other coordinate in this system is the *azimuth* ψ which is zero at solar noon and increases towards the east. During the daily motion, the solar declination δ can usually be assumed constant and equal to its value at midday.

The following relationships will be needed in the next section to calculate the amount of solar radiation falling on an inclined surface:

$$\sin\alpha = \sin\delta \, \sin\phi + \cos\delta \, \cos\phi \, \cos\omega = \cos\theta_z \qquad (2.2)$$
$$\cos\psi = (\sin\alpha \, \sin\phi - \sin\delta)/\cos\alpha \, \cos\phi \qquad (2.3)$$

These equations can be used to determine the sunrise hour angle ω_s:

$$\omega_s = \cos^{-1}(-\tan\phi \, \tan\delta) \qquad (2.4)$$

The sunset hour angle is then simply $-\omega_s$.

2.2.3 Radiation on an inclined surface

The solar radiation data, when available, are frequently given in the form of *global radiation on a horizontal surface* at the site, for example, in the form of global daily irradiation values (denoted by G) for a typical day in any one month (see, for example, Palz, 1984, and Lof *et al.*, 1966). Since PV panels are usually positioned at an angle to the horizontal plane, the energy input to the system must be calculated from the data.

The block diagram of a standard calculation to determine the daily radiation on an inclined surface is shown in Fig. 2.8. The calculation proceeds in three steps. In the first step, the data for the site are used to determine the separate diffuse and beam contributions to the global irradiation on the horizontal plane. This is done by using the extraterrestrial daily irradiation B_0 as a reference, and calculating the ratio $K_T = G/B_0$, known as the *clearness index*. K_T describes the (average) attenuation of solar radiation by the atmosphere at a given site during a given month. In the evaluation of B_0, the variation of the extraterrestrial irradiance on account of the

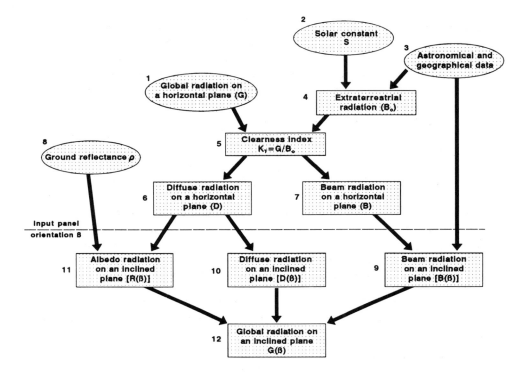

Fig. 2.8 Global radiation on an inclined plane

eccentricity of the Earth's orbit (approximately ± 3%) is usually also taken into account.

In the second step, the diffuse irradiation is obtained using the empirical rule that the diffuse fraction D/G of the global radiation is a universal function of the clearness index K_T. Since $B = G - D$, this procedure determines both the diffuse and beam irradiation on the horizontal plane.

In the third step, the appropriate angular dependences of each component are used to determine the diffuse and beam irradiation on the inclined surface. With allowance for the reflectivity of the surrounding area, the albedo can also be determined. The total daily irradiation on the inclined surface is then obtained by adding the three contributions.

As we shall further discuss in Chapter 4, the daily irradiation values on an inclined panel for each month of the year play an important part in the design of a photovoltaic system.

In more detail now, the calculation proceeds as follows.

The inputs to the calculation are the global daily irradiation G for a day near the middle of each month of the year [block 1], the solar constant S

and its variation due to the eccentricity of the Earth's orbit [block 2], the site's geographical latitude and the solar declination δ for the days of the year in question [block 3]. During the calculation, one also has to input the panel inclination β and the reflectivity ρ of the surrounding area [block 8]. Some typical values of surface reflectivity are given in Table 2.1.

Table 2.1 Typical reflectivity values for some ground covers. (Note: This table is presented for illustration only, as a considerable range of values is encountered for each of the materials. For details, the student should consult the bibliography of this chapter)

Ground cover	Reflectivity
Dry bare ground	0.2
Dry grassland	0.3
Desert sand	0.4
Snow	0.5 – 0.8

[4] The radiation B_o received, over one day, by a unit horizontal area outside the Earth's atmosphere is calculated using the expression

$$B_o = \frac{24}{\pi} S\{1 + 0.33 \cos(2\pi d_n/365)\} (\cos\phi \cos\delta \sin\omega_s + \omega_s \sin\phi \sin\delta)$$

where d_n is the number of the day in the year (1 on 1st January and 365 on 31st December).

[5] The clearness index K_T is now calculated for each month of the year from

$$K_T = G/B_o$$

[6, 7] Various empirical formulas are available for the calculation of the diffuse radiation. We shall adopt here the formula due to Page (1961) which is very simple and accurate enough for the present purposes:

$$D/G = 1 - 1.13 K_T$$

The beam irradiation is then calculated from

$$B = G - D$$

[9] The beam irradiance $B(\beta)$ on a south-facing panel inclined at an angle β to the horizontal surface is now given by

$$B(\beta) = B \frac{\cos(\phi - \beta) \cos\delta \sin[\omega_s(\phi - \beta)] + \omega_s(\phi - \beta) \sin(\phi - \beta) \sin\delta}{\cos\phi \cos\delta \sin\omega_s + \omega_s \sin\phi \sin\delta} \quad (2.5)$$

[10] Assuming that the diffuse radiation is distributed isotropically over the sky dome, the diffuse radiation on the inclined surface is given by

$$D(\beta) = \tfrac{1}{2}(1 + \cos\beta)D \tag{2.6}$$

[8, 11] The albedo radiation. The irradiance reflected from the ground is generally small, and a simple isotropic model is usually sufficient. This gives the result

$$R(\beta) = \tfrac{1}{2}(1 - \cos\beta)\rho D \tag{2.7}$$

The albedo radiation takes on special importance for photovoltaic modules (called *bifacial*) which can utilise energy incident both from the front and from the rear of the panel.

[12] We can now find the total global irradiation $G(\beta)$ on the inclined surface as a sum of the all the contributions (5)–(7):

$$G(\beta) = B(\beta) + D(\beta) + R(\beta)$$

As an illustration, Table 2.2 shows the data for different panel orientations obtained for Barcelona on the Mediterranean coast of Spain. These data will be used in Chapter 4 to discuss the detailed design of a PV system. Presently, we shall consider a few general rules regarding the orientation of the PV modules to capture the optimum amount of solar radiation.

Table 2.2 Daily irradiation in Barcelona (in kWh/m^2) for a typical day in every month as a function of the panel inclination in degrees

Angle	Jan	Feb	Mar	Apr	May	Jun	Jul	Aug	Sep	Oct	Nov	Dec	Annual
0	2.25	2.92	3.88	4.98	5.66	6.38	6.67	5.75	4.20	3.18	2.23	1.70	4.15
5	2.56	3.19	4.08	5.09	5.69	6.36	6.68	5.84	4.46	3.42	2.49	1.93	4.32
10	2.86	3.44	4.26	5.18	5.70	6.32	6.66	5.90	4.61	3.64	2.74	2.16	4.46
15	3.14	3.67	4.42	5.25	5.67	6.25	6.61	5.94	4.73	3.84	2.98	2.37	4.57
20	3.40	3.87	4.55	5.28	5.62	6.15	6.52	5.94	4.82	4.03	3.19	2.57	4.66
25	3.65	4.05	4.66	5.29	5.54	6.03	6.41	5.91	4.89	4.18	3.39	2.75	4.73
30	3.86	4.21	4.73	5.26	5.44	5.88	6.26	5.85	4.93	4.31	3.57	2.92	4.77
35	4.05	4.34	4.78	5.21	5.31	5.70	6.08	5.75	4.94	4.42	3.72	3.07	4.78
40	4.22	4.45	4.81	5.13	5.15	5.49	5.88	5.63	4.93	4.50	3.85	3.19	4.77
45	4.36	4.53	4.80	5.03	4.97	5.26	5.65	5.48	4.88	4.55	3.96	3.30	4.73
50	4.47	4.58	4.77	4.89	4.77	5.01	5.39	5.29	4.81	4.57	4.04	3.39	4.66
55	4.55	4.60	4.71	4.73	4.55	4.74	5.11	5.09	4.71	4.57	4.09	3.45	4.57
60	4.60	4.59	4.62	4.55	4.30	4.45	4.80	4.85	4.58	4.53	4.12	3.49	4.46
65	4.62	4.55	4.50	4.34	4.04	4.14	4.48	4.59	4.42	4.47	4.12	3.51	4.32
70	4.61	4.49	4.36	4.11	3.77	3.83	4.15	4.31	4.25	4.38	4.10	3.51	4.15
75	4.57	4.39	4.19	3.86	3.48	3.50	3.80	4.02	4.05	4.27	4.04	3.48	3.97
80	4.50	4.27	4.00	3.59	3.18	3.17	3.44	3.70	3.82	4.13	3.97	3.43	3.77
85	4.40	4.13	3.79	3.31	2.88	2.84	3.08	3.37	3.58	3.96	3.87	3.36	3.55
90	4.27	3.95	3.55	3.02	2.57	2.51	1.86	3.04	3.32	3.78	3.74	3.27	3.24

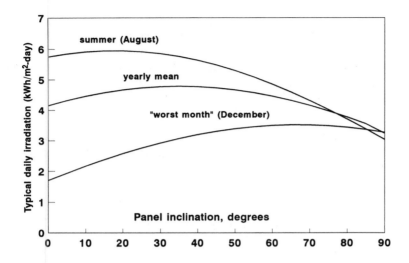

Fig. 2.9 Daily irradiation in Barcelona as a function of panel inclination

To this end, Fig. 2.9 shows the mean daily irradiation in Barcelona during different parts of the year (summer, winter, and yearly average) as a function of the panel inclination. This will allow us to select the optimum inclination for some typical applications.

In some systems (for example, those connected to the grid), the paramount consideration is to collect the maximum energy over the year. It can be seen that the optimum panel angle which maximises the yearly average irradiation lies between 30° and 40°. In general, it is usually close to the latitude angle.

In many stand-alone systems which rely on energy storage by batteries, the principal consideration may not be the total energy received but the daily irradiation during the month with the least sunshine (here, December). In this instance, one should choose an angle between 60° and 70°.

Solar electricity is best suited for applications where peak consumption occurs during the summer (for example, crop irrigation). Here, a panel angle close to 20° would be the best option.

Fig. 2.10 shows the typical daily irradiation in each month of the year for the three angles 20°, 40° and 60° which could be chosen in the above three applications.

SUMMARY OF THE CHAPTER

The nature of solar radiation on Earth was reviewed, and the principal components of the radiation were introduced. We have seen how the solar radiation, both in magnitude and structure, depends on the geographical location. The astronomical relationship between the Earth and the Sun was discussed, defining the principal angles which determine the apparent motion of the Sun. These results were used to evaluate the daily solar irradiation on an inclined surface which represents the energy input to a photovoltaic system.

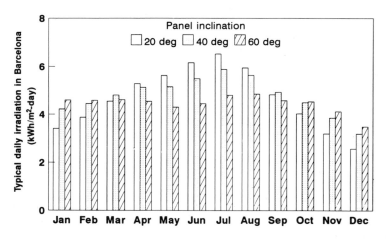

Fig. 2.10 Daily irradiation in Barcelona over the year for selected angles of panel inclination

We have seen that inclination close to the latitude angle of the site will maximise the radiation on an inclined panel over the whole year. In stand-alone systems, it is usual to choose a somewhat steeper angle to minimise storage requirements. For a summer peaking load, on the other hand, a more shallow angle is usually preferred.

BIBLIOGRAPHY AND REFERENCES

IQBAL, M., *An Introduction to Solar Radiation*, Academic, New York, 1983.

LOF, G. O. F., DUFFIE, F. A. and SMITH, C. O., *World Distribution of Solar Radiation*, University of Wisconsin Report No. 21, 1966.

LORENZO, E., Solar Radiation, *in:* Luque A., *Solar Cells and Optics for Photovoltaic Concentration*, Adam Hilger, Bristol, 1989, pp. 268–304.

PAGE, J. K., The estimation of monthly mean values of daily total short-wave radiation on vertical and inclined surfaces from sunshine records for latitudes 40°N–40°S, *in: Proc. United Nations on New Sources of Energy*, Vol. 4, 1961, pp. 378–390.

PALZ, W., ed. *European Solar Radiation Atlas*, Volumes 1 and 2, 2nd edn, Verlag TUV Rheinland, Cologne, 1984.

SELF-ASSESSMENT QUESTIONS

1 What is the approximate solar irradiance on a clear day with the Sun at zenith?
   (a) 1 W/cm^2; (b) 0.01 W/cm^2; (c) 0.1 W/cm^2.

2 What are the three components of solar radiation on an inclined panel?

3 What are the main physical processes that determine the attenuation of solar radiation in the atmosphere?

4 Which of the following statements are true?
   (a) The average daily irradiation over the year is maximum at the equator.
   (b) Regions with high latitude experience relatively high amounts of diffuse radiation.
   (c) The seasonal variation of daily solar irradiation is approximately independent of the latitude.

5 What are the main angular quantities that characterise the (a) yearly and (b) daily apparent motion of the Sun?

6 What is the maximum solar elevation at a site with with latitude 40°N and when does it occur?

7 What measured solar radiation data are usually used to determine the energy input to the PV system?

8 Describe the principal steps of the calculation to determine the solar radiation on an inclined surface.

9 For a photovoltaic installation at latitude 30°N, you have the option to position the panels at an angle 0°, 10°, 30°, 50°, or 90° to the horizontal plane. Which panel orientation would you choose to obtain the maximum energy: (a) over the year; (b) in the summer; (c) in the winter?

Answers

1 (c).
2 Direct (beam), diffuse and surface albedo.
3 Scattering and absorption.
4 (a) F; (b) T; (c) F (see end of section 1).
5 (a) Solar declination.
   (b) Hour angle or azimuth, and the solar elevation or zenith angle.
6 90° − (40° − 23.5°) = 73.5°; at solar noon on the day of summer equinox (21/22 June).
7 Global radiation on a horizontal surface.
8 See figure 2.8.
9 (a) 30°; (b) 10°; (c) 50°.

3

Solar Cells

AIMS

The aims of this unit are to explain how solar cells work and how they are manufactured.

OBJECTIVES

When you complete this unit you should be able to:

1. Analyse the structure of a solar cell.
2. Explain the power output from a cell in terms of the incident energy flux and the electronic structure of the semiconductor.
3. Evaluate the cell performance by using its current–voltage characteristic.
4. Assess the operation of practical devices and limits imposed on their performance.
5. Identify the technological steps which are used in the manufacture of solar cells.

NOTATION AND UNITS

Symbol		SI unit	Other unit
A	Surface area of the cell	m^2	
c	Speed of light in vacuum	m/s	
E_c	Energy of the bottom of the conduction band	J	eV
E_v	Energy of the top of valence band	J	eV

Written by A. Cuevas (Silicon solar cell technology), R. Hill (Thin-film solar cells) and T. Markvart.

(*Contd.*)

Symbol		SI unit	Other unit
E_g	Energy gap	J	eV
$E_{ph}(\lambda)$	Photon energy	J	eV
FF	Fill factor		
G	Irradiance	W/m^2	
h	Planck constant	J s	
I	Electric current	A	
I_{sc}	Short-circuit current	A	
I_ℓ	Light-generated current	A	
I_m	Current at maximum power	A	
I_o	Dark saturation current	A	
J	Electric current density	A/m^2	
k	Boltzmann constant	J/K	eV/K
m	Nonideality factor		
$\mathcal{N}$	Photon flux (density)	m^{-2}s^{-1}	
P_{max}	Power produced by the cell at the maximum power point	W	
q	Electron charge	C	
R_s	Series resistance	Ω	
T	Absolute temperature	K	
V	Voltage	V	
V_{OC}	Open-circuit voltage	V	
V_m	Voltage at maximum power	V	
η	Cell efficiency		
λ	Wavelength of light	m	
	Spectral flux density	m^{-2}s^{-1}eV^{-1} *or* m^{-2}s$^{-1}\mu$m^{-1}	

Unit conversion factors

$1 \text{ eV} = 1.602 \times 10^{-19} \text{ J}$

Values of physical constants

$k = 1.38 \times 10^{-23} \text{ J/K} = 86.3 \times 10^{-3} \text{ eV/K}$
$q = 1.602 \times 10^{-19} \text{ C}$
$c = 2.998 \times 10^{8} \text{ m/s}$
$h = 6.626 \times 10^{-34} \text{ J s}$

3.1 INTRODUCTION

In this chapter, you will learn how solar cells work and how they are manufactured.

A broad overview of the available solar cells will be given in section 3.2.

You will then be shown in section 3.3 how solar cells work. Starting with the electronic properties of semiconductors we shall examine the over-all structure of the cell and explain how it produces electricity. Various practical devices will be discussed, together with power losses which occur during their operation.

The devices that you are most likely to encounter—the silicon and thin-film solar cells—will then be analysed in detail in sections 3.4 and 3.5. You will be shown how they are manufactured, from raw materials to the final device. Emphasis will be placed not only on the electrical properties of the device, but also on the economic aspects of the technology.

3.2 WHAT ARE SOLAR CELLS?

Solar cells represent the fundamental power conversion unit of a photovoltaic system. They are made from semiconductors, and have much in common with other solid-state electronic devices, such as diodes, transistors and integrated circuits. For practical operation, solar cells are usually assembled into modules.

Many different solar cells are now available on the market, and yet more are under development. The range of solar cells spans different materials and different structures in the quest to extract maximum power from the device while keeping the cost to a minimum. Devices with efficiency exceeding 30% have been demonstrated in the laboratory. The efficiency of commercial devices, however, is usually less than half this value.

Crystalline silicon cells hold the largest part of the market. To reduce the cost, these cells are now often made from multicrystalline material, rather than from the more expensive single crystals. Crystalline silicon cell technology is well established. The modules have a long lifetime (20 years or more) and their best production efficiency is approaching 18%.

Cheaper (but also less efficient) types of silicon cells, made in the form of amorphous thin films, are used to power a variety of consumer products. You will be familiar with the solar-powered watches and calculators, but larger amorphous silicon solar modules are also available.

A variety of compound semiconductors can also be used to manufacture thin-film cells (for example, cadmium telluride or copper indium diselenide). These modules are now beginning to appear on the market and hold the promise of combining low cost with acceptable conversion efficiencies.

A particular class of high-efficiency solar cells from single crystal silicon or compound semiconductors (for example, gallium arsenide or indium phosphide) are used in specialised applications, such as to power satellites or in systems which operate under high-intensity concentrated sunlight. The operation and applications of these devices will be reviewed in Chapters 5 and 7.

Summary

Various types of solar cells have been introduced: crystalline and amorphous silicon cells, compound thin-film devices, and high-efficiency cells for specialised applications.

3.3   HOW SOLAR CELLS WORK

3.3.1   Introduction

The solar cell operation is based on the ability of semiconductors to convert sunlight directly into electricity by exploiting the *photovoltaic effect*. In the conversion process, the incident energy of light creates mobile charged particles in the semiconductor which are then separated by the device structure and produce electrical current.

The electronic structure of semiconductors will be reviewed in section 3.3.2. You will learn here about the characteristic distribution of electron energies within the semiconductor, and how the electrical properties of semiconductors can be controlled by the addition of impurities. We shall introduce an essential part of the solar cell, the semiconductor junction, and show how the illumination creates mobile charged particles, electrons and holes. The electrical characteristics of the solar cell will then be obtained by analysing the charge currents across the junction.

In section 3.3.4, various device structures will be reviewed and we show how the device design aims to minimise the losses of power which occur during the cell operation. We shall also examine how the operation of the cell is affected by practical operating conditions, particularly by variable temperature and irradiance.

3.3.2   Electronic structure of semiconductors

3.3.2.1   Band structure, doping

The principles of semiconductor physics are best illustrated by the example of silicon, a group 4 elemental semiconductor. The silicon crystal forms the so-called *diamond lattice* where each atom has four nearest neighbours at the vertices of a tetrahedron. The four-fold tetrahedral coordination is the result of the bonding arrangement which uses the four outer (valence) electrons of each silicon atom (Fig. 3.1). Each bond contains two electrons, and you can easily see that all the valence electrons are taken up by the bonds. Most other industrially important semiconductors crystallise in closely related lattices, and have a similar arrangement of the bonding orbitals.

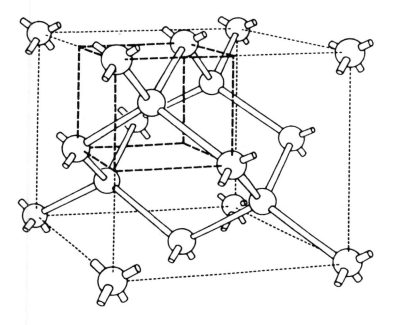

Fig. 3.1 The diamond lattice

This crystal structure has a profound effect on the electronic and optical properties of the semiconductor.

Table 3.1 Energy gaps of principal semiconductors for photovoltaic applications (gap values given at room temperature)

Material	Energy gap (eV)	Type of gap
crystalline Si	1.12	indirect
amorphous Si	~ 1.75	direct
$CuInSe_2$	1.05	direct
CdTe	1.45	direct
GaAs	1.42	direct
InP	1.34	direct

According to the quantum theory, the energy of an electron in the crystal must fall within well-defined *bands*. The energies of valence orbitals which form bonds between the atoms represent just such a band of states, the *valence band*. The next higher band is the *conduction band* which is separated from the valence band by the *energy gap*, or *bandgap*. The width of the bandgap $E_c - E_v$ is a very important characteristic of the semiconductor and is usually denoted by E_g. Table 3.1 gives the bandgaps of the most important semiconductors for solar-cell applications.

A pure semiconductor (which is called *intrinsic*) contains just the right number of electrons to fill the valence band, and the conduction band is therefore empty (Fig. 3.2). Electrons in the full valence band cannot move—just as, for example, marbles in a full box with a lid on top. For practical purposes, a pure semiconductor is therefore an insulator.

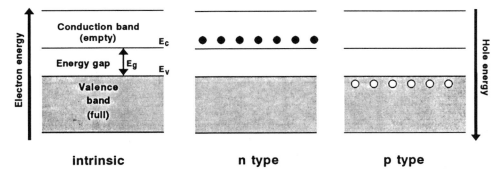

Fig. 3.2 Band diagram and electron–hole distribution in semiconductors

Semiconductors can only conduct electricity if carriers are introduced into the conduction band or removed from the valence band. One way of doing this is by alloying the semiconductor with an impurity. This process is called *doping*. As we shall see, doping makes it possible to exert a great deal of control over the electronic properties of a semiconductor, and lies at the heart of the manufacturing process of all semiconductor devices.

Suppose that some group 5 impurity atoms (for example, phosphorus) are added to the silicon melt from which the crystal is grown. Four of the five outer electrons are used to fill the valence band and the one extra electron from each impurity atom is therefore promoted to the conduction band (Fig. 3.2). For this reason, these impurity atoms are called *donors*. The electrons in the conduction band are mobile, and the crystal becomes a conductor. Since the current is carried by negatively charged electrons, this type of semiconductor is called *n-type*.

A similar situation occurs when silicon is doped with group 3 impurity atoms (for example, boron) which are called *acceptors*. Since four electrons per atom are needed to fill the valence band completely, this doping creates electron deficiency in this band (Fig. 3.2). The missing electrons—called *holes*—behave as positively charged particles which are mobile, and carry current. A semiconductor where the electric current is carried predominantly by holes is called *p-type*.

The prevailing charge carriers in a given semiconductor are called *majority carriers*. Examples of majority carriers are electrons in an n-type semiconductor and holes in the p-type. The opposite type of carriers whose concentration is generally much lower, are called *minority carriers*.

3.3.2.2 Semiconductor junctions

The operation of solar cells is based on the formation of a *junction*. Various examples of junctions are shown in Fig. 3.3. Perhaps the simplest is the *p-n*

Fig. 3.3 Band diagram of semiconductor junctions

junction, an interface between the n and p regions of one semiconductor. A layer of intrinsic material is sometimes incorporated between the n- and p-type regions, resulting in a wider transition zone. In contract with these *homojunctions*, a *heterojunction* is formed by two different semiconductors— note the difference in the bandgaps on the two sides of the junction.

An interface between a metal and a semiconductor may also form a junction, called the *Schottky barrier*. In general, the properties of metal contacts with a semiconductor depend on the actual materials in question. For each semiconductor, some metals form a Schottky barrier but some form an *ohmic contact* where the barrier is absent. These contacts are used to extract electrical current from the device.

The important feature of all junctions is that they contain a strong electric field. To illustrate how this field comes about, let us imagine the hypothetical situation where the p-n junction is formed by joining together two pieces of semiconductor, one p-type and the other n-type (although this manner of junction formation is not normally employed in practice, it is a convenient means to demonstrate the relevant principles). In separation, there is electron surplus in the n-type material and hole surplus in the p-type. When the two pieces are brought into contact, electrons from the n region near the interface diffuse into the p side, leaving behind a layer which is positively charged by the donors. Similarly, holes diffuse in the opposite direction, leaving behind a negatively charged layer stripped of holes. The resulting junction region then contains practically no mobile charge carriers (Fig. 3.4), and the fixed charges of the dopant atoms create a potential barrier acting against a further flow of electrons and holes. Note that the electric field in the junction pulls the electrons and holes in opposite directions.

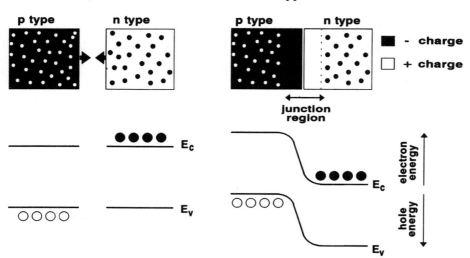

Fig. 3.4 Diagram of p-n junction formation and the resulting band structure

Fig. 3.5 shows the band diagram of a p-n junction diode in equilibrium, and when external voltage is connected to the diode. Without bias, of course, there is no current through the junction. We may imagine this zero net current as consisting of two very small opposite currents I_o and $-I_o$ which remain from the current flow prior to the junction formation. These currents are very small indeed, corresponding to a current density of the order of 10^{-14} A/cm^2 in a good silicon diode.

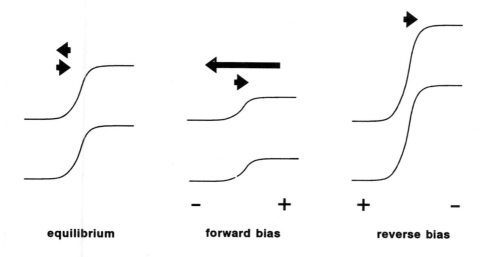

equilibrium   **forward bias**   **reverse bias**

Fig. 3.5 Electric currents in a p-n junction under external bias

This current balance is altered considerably when voltage is applied to the junction. A forward bias, i.e. positive voltage applied to the p side, reduces the height of the potential barrier. This, in turn, increases dramatically the current through the diode.

Under reverse bias, on the other hand, the barrier is increased. This has a much smaller effect on the device, and produces only the tiny current I_o (the *dark saturation current*)—which is much smaller than the current under forward bias. The junction therefore acts as a rectifier, or diode.

In mathematical terms, the I–V characteristic of a diode is given by the Shockley equation

$$I = I_o \left[\exp\left(\frac{qV}{kT}\right) - 1 \right] \tag{3.1}$$

where I is the current, V is the voltage, k is the Boltzmann constant, q is the magnitude of the electron charge, and T is the absolute temperature. The I–V characteristic (3.1) is shown in Fig. 3.6.

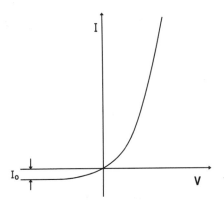

Fig. 3.6 The diode *I–V* characteristic

3.3.2.3 Light absorption by a semiconductor

Photovoltaic energy conversion relies on the quantum nature of light whereby we perceive light as a flux of particles—*photons*—which carry the energy

$$E_{ph}(\lambda) = \frac{hc}{\lambda} \tag{3.2}$$

where h is the Planck constant, c is the speed of light, and λ is the wavelength of light. On a clear day, about 4.4×10^{17} photons strike a square centimetre of the Earth's surface every second.

Only some of these photons—those with energy in excess of the bandgap—can be converted into electricity by the solar cell. When such a photon enters the semiconductor, it may be absorbed and promote an electron from the valence to the conduction band (Fig. 3.7). Since a hole is left behind in the valence band, the absorption process generates *electron–hole pairs*.

Each semiconductor is therefore restricted to converting only a part of the solar spectrum (Fig. 3.8). Using equation (3.2), the spectrum has been plotted here in terms of the incident photon flux as a function of photon energy. The shaded area represents the photon flux that can be converted by a silicon cell—about two thirds of the total flux.

The nature of the absorption process also indicates how a part of the incident photon energy is lost in the event. Indeed, it is seen that practically all the generated electron–hole pairs have energy in excess of the bandgap. Immediately after their creation, the electron and hole decay to states near the edges of their respective bands. The excess energy is lost as heat and cannot be converted into useful power. This represents one of the fundamental loss mechanisms in a solar cell.

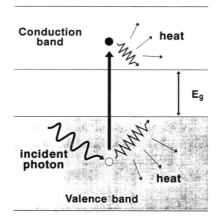

Fig. 3.7 The generation of electron–hole pairs by light

Fig. 3.8 Photon flux utilised by a silicon solar cell

We can make some rough estimates of the magnitude of electrical power that can be produced. To this end, let us interpret the light-induced electron traffic across the bandgap as electron current, called the *generation current*. We shall see shortly (section 3.3.3) that a solar cell can indeed transform this fictitious current into real electric current across the device. Neglecting losses, each photon then contributes one electron charge to the generation current. The electric current is then equal to

$$I_\ell = q \mathcal{N} A \tag{3.3}$$

where $\mathcal{N}$ is the number of photons in the highlighted area of the spectrum, and A is the surface area of the semiconductor that is exposed to light. For example, the current density $J_\ell = I_\ell/A$ that corresponds to the terrestrial spectrum is about $1.6 \times 10^{-19} \times 4.4 \times 10^{17} = 70$ mA/cm². Of this, a silicon solar cell can convert at most 44 mA/cm².

What voltage can a solar cell generate? One can obtain an upper bound by a simple electrostatic argument. As we shall see, the electric power is produced by separating the light-generated electrons and holes to the terminals of the device. This separation can only proceed if the electrostatic energy of the charges after separation (qV where V is the voltage at the terminals) does not exceed the pair energy in the semiconductor, equal to the bandgap. This sets an upper limit on the voltage in the form

$$V = E_g/q. \tag{3.4}$$

The *maximum voltage* in volts is thus numerically equal to the bandgap of the semiconductor in electronvolts. Although the actual voltage achieved in practice is considerably less than the theoretical limit, the trend expressed by (3.4) that wide-bandgap semiconductors produce higher voltage generally holds true.

We have so far assumed that all the photons with energy in excess of the bandgap are absorbed. Indeed, many semiconductors are good light absorbers, and absorb all the above-bandgap light in a layer of few micrometres thick. They are called *direct-bandgap semiconductors*. In others—*indirect-gap semiconductors* which include also crystalline silicon—the absorption process is more complicated. A quantum of lattice vibrations must participate in the conversion of a photon into an electron—hole pair to conserve momentum which hinders the process and decreases the capability of the semiconductor to absorb light. This phenomenon is illustrated in Fig. 3.9. Note that several hundred micrometres of silicon are necessary to absorb all the above-bandgap light but few micrometres of a direct-gap material (for example, GaAs) are sufficient for this purpose.

Fig. 3.9 Optical absorption properties of direct- and indirect-gap semiconductors

3.3.3 The solar cell

Fig. 3.10 shows the diagram of a silicon cell, the typical solar cell in use today. The electrical current generated in the semiconductor is extracted by contacts to the front and rear of the cell. The top contact structure which must allow light to pass through is made in the form of widely-spaced thin metal strips (usually called *fingers*) that supply current to a larger bus bar. The cell is covered with a thin layer of dielectric material—the *antireflection coating* or ARC—to minimise light reflection from the top surface.

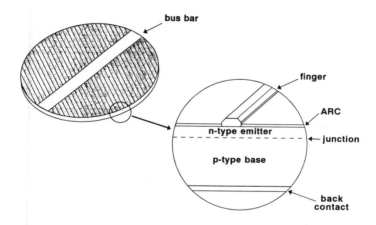

Fig. 3.10 The silicon solar cell

Fig.3.11 shows the band diagram of the semiconductor section under illumination. Light generates electron–hole pairs on both sides of the junction, in the n-type emitter and in the p-type base. The generated minority carriers—electrons from the base and holes from the emitter—then diffuse to the junction and are swept away by the electric field, thus producing electric current across the device. Note how the electric currents of the electrons and holes reinforce each other since these particles carry opposite charges. The p-n junction therefore separates the carriers with opposite charge, and transforms the generation current I_ℓ between the bands into an electric current across the p-n junction.

The I–V characteristic of a solar cell can be obtained by drawing an equivalent circuit of the device (Fig. 3.12). The generation of current I_ℓ by light is represented by a current generator in parallel with a diode which represents the p-n junction. The output current I is then equal to the difference between the light-generated current I_ℓ and the diode current I_D. Equation (3.1) then gives

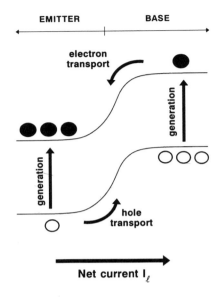

Fig. 3.11 Currents in a p-n junction under illumination

$$I = I_\ell - I_o \left[\exp\left(\frac{qV}{kT}\right) - 1 \right]. \tag{3.5}$$

Note that, under open circuit when $I = 0$, all the light-generated current passes through the diode. Under short circuit ($V = 0$) on the other hand, all this current passes through the external load. The I–V characteristic (3.5) and its relationship to the diode characteristic are shown in Fig. 3.12.

The I–V characteristic contains several important points. One is the short-circuit current I_{sc} which, as we noted, is simply the light-generated current I_ℓ. The second is the open-circuit voltage V_{oc} obtained by setting $I = 0$:

$$V_{oc} = \frac{kT}{q} \ln\left(\frac{I_\ell}{I_o} + 1\right). \tag{3.6}$$

It is worthwhile to examine this equation in more detail. Both I_ℓ and I_o depend on the structure of the device. However, it is the value of I_o—which can vary by many orders of magnitude, depending on the device geometry and processing—that determines the open circuit voltage in practical devices.

No power is generated under short or open circuit. The maximum power P_{max} produced by the device is reached at a point on the characteristic where the product IV is maximum. This is shown graphically in Fig. 3.13 where the position of the *maximum power point* represents the largest area of the rectangle shown. One usually defines the fill-factor FF by

$$P_{max} = V_m I_m = FF \ V_{oc} I_{sc} \qquad (3.7)$$

where V_m and I_m are the voltage and current at the maximum power point.

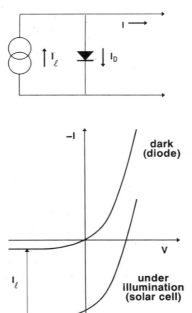

Fig. 3.12   The equivalent circuit and *I–V* characteristic of a solar cell compared to a diode

Fig. 3.13   The *I–V* characteristic of a solar cell with the maximum power point

The *efficiency* η of a solar cell is defined as the power P_{max} produced by the cell at the maximum power point under standard test conditions, divided by the power of the radiation incident upon it. Most frequent conditions are: irradiance $100 \, mW/cm^2$, standard reference AM1.5 spectrum, and temperature $25°C$. The use of this standard irradiance value is particularly convenient since the cell efficiency in percent is then numerically equal to the power output from the cell in mW/cm^2. Other test conditions are sometimes adopted for cells which operate in a different environment, for example, cells which power satellites and operate under AM0 spectrum.

3.3.4 Typical device structures

The *I–V* characteristic (3.5) which we have derived for a simplified device describes, in fact, rather well the operation of solar cells in practice if the parameters make allowance for the losses which take place in practical devices. The typical production silicon cell (Fig. 3.10) has already been mentioned. Some modern types of research silicon cells are shown in Fig. 3.14. The *passivated emitter solar cell* (PESC) structure has been developed at the University of New South Wales in Australia for operation

Fig. 3.14 The structure and band diagram of high-efficiency silicon solar cells

under ordinary sunlight. The *point-contact cell* of Stanford University, USA, has been designed for optimum operation under concentrated sunlight. Gallium arsenide solar cells (Fig. 3.15) are, because of their high cost, usually intended for operation on satellites or in concentration systems. The typical amorphous silicon solar-cell structure (Fig. 3.16) and the tandem cell (Fig. 3.17) will be discussed in detail below and in section 3.5.

Fig. 3.15 The structure and band diagram of gallium arsenide solar cells

Fig. 3.16 The structure of amorphous silicon p-i-n solar cell

Fig. 3.17 The structure and spectral contributions of the tandem cell

3.3.4.1 *Power losses in solar cells*

Fundamental losses. As we have seen in section 3.3.2.3, carrier generation in the semiconductor by light involves considerable dissipation of the generated carrier energy into heat. In addition, a considerable part of the solar spectrum is not utilised because of the inability of a semiconductor to absorb the below-bandgap light.

Can these losses be reduced? Yes, but not with a simple structure that we have in mind at the moment. Such a device is called a *tandem cell* (Fig. 3.17), and represents a stack of several cells, each operating according to the principles that we have described. The top cell must be made of a high-bandgap semiconductor, and converts the short-wavelength radiation. The transmitted light is then converted by the bottom cell. This arrangement increases considerably the achievable efficiency. Laboratory devices operating at over 30% have been demonstrated. Most of these structures are at an experimental stage but some thin-film devices of this type may not take long to reach fruition, as we shall see in section 3.5.

Recombination. An opposite process to carrier generation is recombination when an electron–hole pair is annihilated. Recombination is most common at impurities or defects of the crystal structure, or at the surface of the semiconductor where energy levels may be introduced inside the energy gap. These levels act as stepping stones for the electrons to fall back into the valence band and recombine with holes (Fig. 3.18). An important site of recombination are also the ohmic metal contacts to the semiconductor.

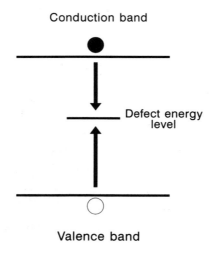

Fig. 3.18   Defect-assisted recombination of electron–hole pairs

What measures can one take to minimise the recombination losses? Surface recombination and recombination at contacts which are considerable in the conventional silicon cell (Fig. 3.10) can be reduced by adapting the device structure. Figs 3.14 and 3.15 show the implementation of such measures in high-efficiency silicon cells. The external surfaces of the semiconductor are here 'protected by a layer of passivating oxide to reduce surface recombination. The top layer of GaAlAs in the GaAs cell has a similar purpose. The contacts are surrounded by heavily-doped regions acting as 'minority-carrier mirrors' which impede the minority carriers from reaching the contacts and recombining.

Recombination reduces both the voltage and current output from the cell. The voltage losses reduce the limiting value (3.4) to the open-circuit voltage (3.6), restricted principally by the dark saturation current I_o. This current contains two contributions, each proportional to the number of recombination events—and therefore the volume—for each region on either side of the junction. This result is exploited in the Stanford point-contact silicon cell (Fig. 3.14) with a greatly reduced emitter volume. A thin silicon cell design which increases the voltage by reducing the base thickness, will be discussed in section 3.5.

The current losses can be grouped under the term of *collection efficiency*, the ratio between the number of carriers generated by light and the number that reaches the junction. Considerations of the collection efficiency affect the design of the solar cell. In crystalline materials, the transport properties are usually good, and carrier transport by simple diffusion is sufficiently effective. In amorphous and polycrystalline thin films, however, electric fields are needed to pull the carriers, as can be seen in the design of the amorphous silicon p-i-n cell (Fig. 3.16).

Other losses to the current produced by the cell arise from light reflection from the top surface, shading of the cell by the top contacts, and incomplete absorption of light. The last feature is particularly significant for crystalline silicon cells since—as we have seen in section 3.3.2.3—silicon has poor light absorption properties. Various measures can be adopted to minimise these losses.

The top contact shading is of particular concern for cells which operate under concentrated, high-intensity sunlight. It is then advantageous to eliminate the shading altogether, as in the Stanford point-contact cell, by moving both contacts to the back of the cell.

The top-surface reflection is usually in the region of 10% for a cell covered with a single-layer ARC, and can be reduced further by the application of two or more layers. Another technique is *surface texturing*. As we shall see in section 3.4, the top surface of a silicon cell can be made in the form of small pyramids which give the reflected light another chance of absorption. When combined with a single-layer ARC, the reflectivity of the surface can be brought down to below 1%.

The absorption of light in silicon cells can be improved by making the back contact optically reflecting. When combined with a textured top surface, this geometry results in effective *light trapping* which provides a good countermeasure for the low absorptivity of silicon. In particular, as we shall see in section 3.5, it allows the manufacture of thin silicon cells.

Series resistance. The transmission of electric current produced by the solar cell involves ohmic losses. These can be grouped together and included as a resistance in the equivalent circuit (Fig. 3.19). It is seen that the series resistance affects the cell operation mainly by reducing the fill factor.

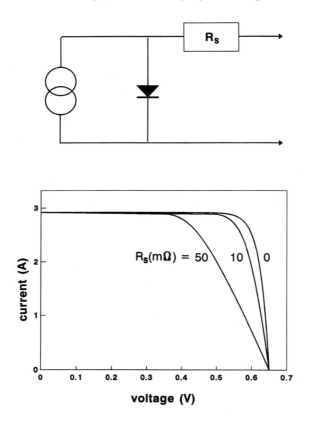

Fig. 3.19 The series resistance and its effect on the *I–V* characteristic of a solar cell

The *I–V* characteristic of a practical device is sometimes best approximated by a modified expression

$$I = I_\ell - I_o \left[\exp\left(\frac{qV + IR_s}{mkT}\right) - 1 \right] \tag{3.8}$$

which also includes the series resistance R_s. Here, m is an empirical *nonideality factor* whose value is usually close to unity. The light-generated current in (3.8) is taken as a measured parameter which takes into account all the current losses in the cell.

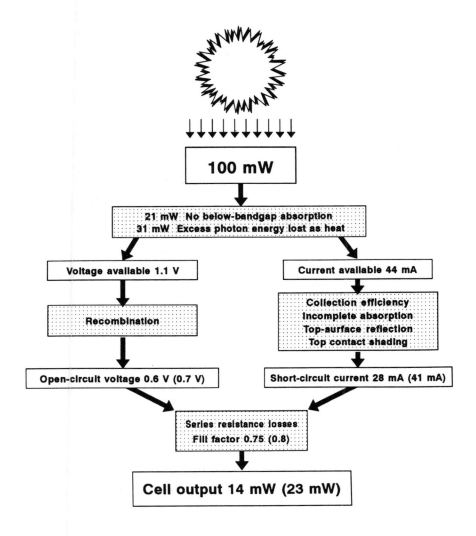

Fig. 3.20 The power losses in silicon solar cells. The figures are per square centimetre for production cells and (in brackets) for laboratory cells

The losses which we have discussed in this section are summarised in Fig. 3.20. It is seen that the fundamental losses reduce the maximum theoretical efficiency of a silicon cell to about 48%. Additional voltage losses (~36%), current losses (~10%), and losses associated with the fill factor (~20%) then explain the efficiency of about 23% for the best silicon cell today. Higher efficiencies have been obtained under concentrated sunlight, in devices from other materials or in tandem structures. The typical production silicon cell has an efficiency of about 14%. However, new devices with efficiency approaching 18% are beginning to appearing on the market.

3.3.4.2 Temperature and irradiance effects

In practical applications, solar cells do not operate under standard conditions. The two most important effects that must be allowed for are are due to the variable temperature and irradiance.

Temperature. This has an important effect on the power output from the cell (Fig. 3.21). The most significant is the temperature dependence of the voltage which decreases with increasing temperature (*its temperature coefficient is negative*). The voltage decrease of a silicon cell is typically 2.3 mV per °C. The temperature variation of the current or the fill factor are less pronounced, and are usually neglected in the PV system design.

Fig. 3.21 Temperature dependence of the *I–V* characteristic of a solar cell

Effects of irradiance. The solar-cell characteristics under different levels of illumination are shown in Fig. 3.22. As we have seen, the light-generated current is proportional to the flux of photons with above-bandgap energy. Increasing the irradiance increases, in the same proportion, the photon flux which, in turn, generates a proportionately higher current. Therefore, the short-circuit current of a solar cell is directly proportional to the irradiance. The voltage variation, as can be seen from equation (3.6), is much smaller (it depends logarithmically on the irradiance), and is usually neglected in practical applications.

Fig. 3.22 Irradiance dependence of the *I–V* characteristic of a solar cell

Summary

The solar cell is a semiconductor device that converts the quantum flux of photons into electric current. When light is absorbed, it first creates electron–hole pairs. These mobile charges are then separated by the electric fields at the junction. The electrical output from the cell is described by the *I–V* characteristic whose parameters can be linked to the material properties of the semiconductor. Various solar-cell structures have been discussed in relation to the principal power losses in a solar cell. In addition to the fundamental losses associated with light absorption, other losses, including recombination and losses dependent on the structure of the device, have been analysed in some detail.

3.4 SILICON SOLAR CELL TECHNOLOGY

3.4.1 Introduction

The technology based on crystalline silicon is the most reliable and most developed photovoltaic technology at the present time. It is not simple, however, and requires the use of sophisticated equipment and complex technological processes. Four major stages need to be followed to make photovoltaic modules from sand:

- from sand to pure silicon;
- growth of silicon crystals;
- from wafers to solar cells;
- from cells to module.

The technology has much in common with the microelectronics industry, and the benefits of the close links have helped to improve the performance of laboratory silicon solar cells to a level not far from the theoretical maximum. However, if photovoltaics is to provide a significant part of the world energy needs, this technology—which has been developed for small devices—must be replaced by new, large-scale and low-cost techniques. In particular, new technologies for the extraction, purification and crystallisation of solar grade silicon are needed. New cell-fabrication procedures have to be adopted by the industry to narrow the gap between the conversion efficiencies obtained in the laboratory and in a commercial process.

An overview of the different steps of silicon cell technology now follows. The conventional technology is described in some detail since it represents not only the past development but, to a large extent, the present state of the industrial production. New techniques that have already made a significant impact on silicon cell manufacture and those still in the research phase are also briefly described since they represent the likely evolution of the technology in the coming years.

3.4.2 From sand to pure silicon

The supply of silicon is practically endless: 60% of the Earth's crust is sand, in major part, quartzite or silicon dioxide (SiO_2). Silicon is produced in large amounts, about 600 000 tonnes a year worldwide, to make special steel and alloys. This *metallurgical-grade silicon* is obtained by reducing quartzite with coke (coal) in electric-arc furnaces. Its purity is only 99.0%—insufficient for electronic applications—but both the energy input (about 50 kWh/kg) and cost (about $2/kg) are relatively low.

The semiconductor industry purifies this feedstock until the impurity concentration is less than 0.2 ppma (parts per million atomic). Since 1 cm³ of crystalline silicon contains 5×10^{22} atoms, this purity implies that the total number of foreign atoms should be less than $10^{16}/cm^3$. The conventional process is described in Fig. 3.23. It includes the chemical transformation of Si into a liquid compound (trichlorosilane, $SiHCl_3$) which allows the desired purity to be reached by distillation. The trichlorosilane vapour then reacts with hydrogen to obtain *semiconductor-grade silicon*. This solidification step takes place in a Siemens reactor at about 1100°C, and is very energy intensive (about 200 kWh/kg). The price of polysilicon thus rises to about $80/kg.

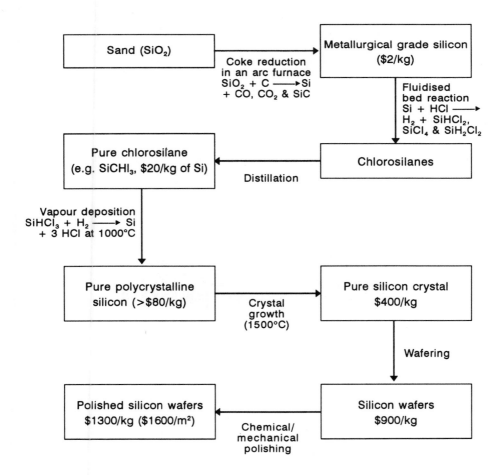

Fig. 3.23 From sand to pure silicon (after T. F. Ciszek, *Proc. Twentieth IEEE Photovoltaic Specialists Conf.*, 1988, p. 31)

Semiconductor grade silicon is, in fact, more pure than is required to make solar cells. The so-called *solar-grade silicon* can contain up to about ten times more impurities (i.e. 1 ppma) and still produce reasonably efficient cells. This facilitates the search for lower cost purification procedures. Several techniques have already been successfully demonstrated on a laboratory scale.

3.4.3 Growth of silicon crystals

Fig. 3.24 shows schematically how a crystal is grown. Silicon is first melted at 1400°C. A small silicon crystal properly cooled is used as a seed to start the crystallisation process. As the seed is pulled out silicon solidifies at the interface with the melt and, if the pulling is slow enough, the silicon atoms arrange themselves according to the crystallographic structure of the seed. This yields an ingot of single-crystal silicon.

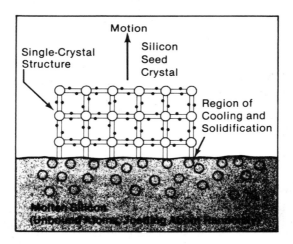

Fig. 3.24 The basics of crystal growth (from P. Hersh and K. Zweibel, *Basic Photovoltaic Principles and Methods*, U.S. Government Printing Office, Washington, DC, 1982. Reprinted with permission)

The degree of purity improves during the growth process since impurities tend to segregate towards the liquid phase. A controlled amount of boron (or phosphorus) is usually added to the melt to dope the silicon p- or n-type.

There are various methods of growing silicon crystals. The one that resembles most closely the basic description is the *Czochralski method* (Fig. 3.25a) which is also the most common in industrial use.

Fig. 3.25 Czochralski (a) and float-zone (b) crystal growth methods (from P. Hersh and K. Zweibel, *Basic Photovoltaic Principles and Methods*, U.S. Government Printing Office, Washington, DC, 1982. Reprinted with permission)

The cylindrical ingots are typically 1 m long, 15 cm in diamter and 40 kg in weight. The growth rate is about 0.1–0.2 cm/min. To increase the through-put, the crucible can be continuously replenished with molten silicon in some machines. Ingots up to 3.5 m long and 150 kg in weight have been grown this way. Another recent development is the use of magnetic fields to reduce the interaction between the molten silicon and the crucible, thus reducing the usual carbon and oxygen contamination from the latter. The state of the art for solar cells made from CZ silicon is 18% efficiency for 100 cm² industrial cells (AEG), and 19% efficiency for a 49 cm² laboratory cell with a laser-grooved metal grid (UNSW).

The purest silicon is obtained by using the *float-zone process*, also shown in Fig. 3.25. The starting polysilicon is first given the shape of a cylindrical bar. The bar is then locally melted by radio frequency induction. By moving the coil, and hence the molten zone, along the bar starting from the seed end, the silicon adopts the crystalline structure. It is significant that the molten zone is self-supporting and is never in contact with a foreign material, thus avoiding contamination problems. A very high purity and also high structural quality can be achieved by performing several passes of the coil. The typical growth rate is 0.3–0.5 cm/min, and the typical ingot is 12.5 cm in diameter and 1 m in length. FZ silicon is the preferred material for the fabrication of high-efficiency cells. The present world record is 23.3% for a 4 cm^2 cell made at the University of New South Wales, Australia.

It is possible to make silicon ingots by pouring molten silicon into a crucible and controlling the cooling rate. This cast silicon does not have the same high quality as the material produced by the CZ or FZ methods, but it is good enough to make solar cells. Rather than being a large single crystal, the ingot is made up of many small crystals, or grains (see Fig. 3.26). For this reason it is called *multicrystalline*. Several manufacturers use slightly different casting methods to produce blocks of 43 × 43 cm^2 cross sectional area and 120 kg in weight. A typical crystallisation rate is 0.5 kg/min. Industrial multicrystalline silicon cells of 100 cm^2 area have attained 15.8% efficiency (Sharp). The laboratory record is 17.8% for a 4 cm^2 cell (UNSW). Commercial cells are about 12–14% efficient.

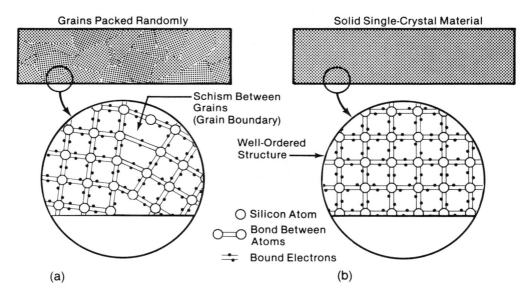

Fig. 3.26 Multicrystalline (a) and monocrystalline (b) silicon (from P. Hersh and K. Zweibel, *Basic Photovoltaic Principles and Methods*, U.S. Government Printing Office, Washington, DC, 1982. Reprinted with permission)

The silicon ingots now have to be sliced into wafers. The usual way to do this in the microelectronic industry is by using inner diameter saws (Fig. 3.27).

Their main disadvantage is that they are very slow (about 30 wafers/h) and that they waste almost half the silicon as *kerf loss* (they yield about 0.7 m² for each kg of ingot). The throughput can be increased if many wafers are cut simultaneously. A multiblade slurry saw machine with 450 cutting blades can produce 2 million wafers per year, that is, 3.6 m²/h. This type of sawing machine can decrease the kerf loss if very thin blades or wires are used. Additional saving can be made by cutting thinner wafers, of thickness 200 μm rather than the customary 450 μm. As an indication, the energy consumed in the crystal growing and slicing processes would decrease from about 740 kWh/m² for the case of cast silicon and inner diameter saws to

Fig. 3.27 Inner diameter saw (a) and multiwire saw (b) (from P. Hersh and K. Zweibel, *Basic Photovoltaic Principles and Methods*, U.S. Government Printing Office, Washington, DC, 1982. Reprinted with permission)

below 300 kWh/m² for the case of multiblade saws and thin wafers. The energy cost of making CZ wafers is about 930 kWh/m².

Considering how cumbersome the sawing of ingots is we should not wonder at the variety of methods that have been investigated to grow the silicon crystal directly in a sheet form. About half a dozen of them are still being intensively developed. A good example is the EFG method (Edge-defined Film-fed Growth), shown in Fig. 3.28. The molten silicon ascends by capillarity through a thin slot in a graphite die. The crystallisation is initiated with a seed that is pulled upwards, thus growing a ribbon several meters long and 300 μm thick. A tubular octagon can be grown using the appropriate die shape, and eight ribbons are subsequently cut using a laser. An interesting feature of most sheet-growth techniques is that they have a high production rate—for example, the rate is 160 cm² for the EFG method. Experimental cells made with EFG silicon have reached 16% (Mobil), although average commercial cells are only about 12% efficient. Other methods have produced experimental cells with efficiencies higher than 10% and very high throughput. This technique is promising but has yet to be further developed.

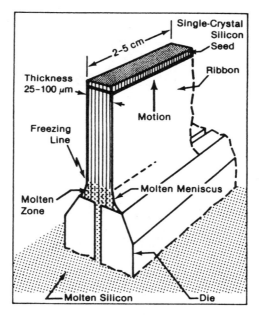

Fig. 3.28 Ribbon growth by +the EFG method (from P. Hersh and K. Zweibel, *Basic Photovoltaic Principles and Methods*, U.S. Government Printing Office, Washington, DC, 1982. Reprinted with permission)

There is one more technique that deserves to be mentioned, if only because it is very different from all the crystallisation methods described so far. By alloying the silicon with a metal (for example, tin) the melting point can be brought down to about 900°C. Under certain conditions, controlled cooling makes the silicon precipitate from the liquid phase on top of a ceramic substrate. The silicon layer thus grown can be arbitrarily thin, although at present a 200 µm layer is being used. Commercial size cells made by Astropower have shown 10% efficiency. We shall discuss this technique further in section 3.5.

Among the total wafer production, cast multicrystalline silicon amounts to about 30% of the wafers for solar cell manufacture. The rest is monocrystalline silicon produced by the Czochralski method. Wafers grown by the ribbon growth technologies have yet to make a significant impact on the market.

3.4.4 Typical cell-fabrication process

To transform a silicon wafer into a solar cell, the wafer is subjected to several chemical, thermal and deposition treatments. The cross section of a silicon cell (Fig. 3.29) shows the different layers that need to be formed. The main steps of the fabrication process are as follows:

1. Surface texturing.
2. p-n junction formation.
3. Possible back p^+ region formation.

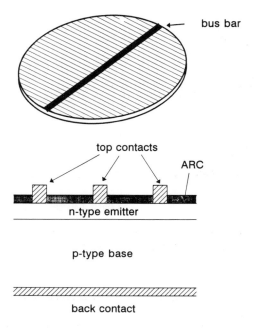

Fig. 3.29 The structure of silicon solar cell

4.   Front and back metal contacts.
5.   Antireflection layer deposition.

The surface of as-cut silicon wafers is heavily damaged by sawing. Chemical etches are then used to remove a few microns of silicon. When monocrystalline silicon is used it is possible to use a second, orientation-dependent etch, to form many small pyramids (Fig. 3.30). The facets of the pyramids reflect light towards other facets, giving it several chances to penetrate into the solar cell. The reflectivity of bare silicon is thus decreased from 33% for a polished surface to 10% for a textured one. Textured surfaces have a black appearance, especially when they are wet or behind a glass, indicating that little of the incident light is reflected.

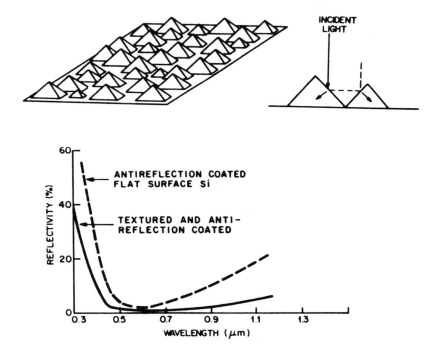

Fig. 3.30 The textured surface and its reflectivity (from P. Hersh and K. Zweibel, *Basic Photovoltaic Principles and Methods*, U.S. Government Printing Office, Washington, DC, 1982. Reprinted with permission)

The wafers are usually p-type. The p-n junction is then formed by thermal diffusion of n-type impurity, usually phosphorus, from the front surface. Phosphorus atoms diffuse into silicon at a temperature of 900°C or higher. Fig. 3.31 shows a quartz diffusion furnace. The dopant source can be a gas (for example, P_2H_2), liquid ($POCl_3$) or a solid (P_2O_5). Alternative techniques include the deposition of phosphorus oxide on the wafers before introducing them into the furnace. This deposition can be done by screen printing, spraying, spinning or by chemical vapour deposition. Ion implantation, a technique quite common in the microelectronics industry, is considered too expensive for photovoltaics, but it has been used in research.

Fig. 3.31 Quartz diffusion furnace

Although it is not absolutely necessary and might be considered irrelevant for low-efficiency cells, a back p^+ region may be formed to improve the cell performance. As we have seen in section 3.3, this feature creates a back surface field that decreases the chances of carriers recombining at the back surface. The easiest way to form it is by depositing an aluminium layer and alloying it at about 800°C, or even diffusing it at about 1000°C.

Electrical contacts are usually formed by screen printing. This technology is inexpensive, simple, and can be automated. The screen consists of a mesh of wires imbedded in an emulsion, as shown in Fig. 3.32. This emulsion is photographically patterned and removed from the places where metal is to be deposited. A paste containing the metal is squeezed through the screen onto the wafer. Upon firing the organic solvents evaporate and the metal powder becomes a conducting path for the electrical current. The paste also contains small amounts of glass that provide a good adhesion to the silicon surface. Although the typical width of the metal grid fingers lies in the range 150–300 μm, a width of 50 μm is within the capabilities of screen printing.

High-efficiency and concentrator cells need much narrower lines to minimise shadowing losses. The 5–10 μm line widths frequently encountered in

Fig. 3.32 Screen printing (from R. J. Van Overstraeten and R. P. Mertens, *Physics, Technology and Use of Photovoltaics*, Adam Hilger, Bristol, 1986, p. 135)

their metal grids are made by photolithography. The metal is deposited in a high-vacuum machine either by heat, electron-beam evaporation, or by sputtering.

A thin layer of a transparent material which acts as *antireflection coating* can be deposited before or after the formation of the metal contacts. This dielectric material has an optimum value of refractive index between those of silicon and glass,

$$n = \sqrt{n_o n_{Si}} = 2.3$$

and the optimum thickness d is given by

$$d = \lambda/4 = 70 \text{ nm}$$

where $n_o = 1.5$ for glass and $n_{Si} = 3.6$ for silicon. Appropriate materials are titanium oxide (TiO_2) and tantalum oxide (Ta_2O_5) with $n = 2$–2.3. Other materials that are used are silicon nitride (Si_3N_4), silicon monoxide (SiO) and aluminium oxide Al_2O_3, with $n = 1.9$. A single-layer antireflection coating can decrease the reflectivity of a flat surface to about 10%, the typical case for multicrystalline silicon cells. Monocrystalline silicon cells are usually textured. In this case, the use of an antireflection coating gives

Table 3.2 Summary of the crystalline silicon cell and module efficiencies

CRYSTALLINE SILICON SOLAR CELLS, AM1.5G				
Material	Cell	Organisation	Area (cm^2)	η (%)
FZ	PERL	UNSW	4	23.3
	Back PCC	Stanford	37.5	22.7
	Simpl. BCC	Stanford	10	21.2
	BCSC	UNSW	12	21.3
CZ	BCSC	UNSW	47	18.3
	$n^+ pp^+$	Telefunken	113	17.3
	$n^+ pp^+$	Sharp	100	17.3
	commercial	several	100	13.0
Multicrystalline	PESC	UNSW	4	17.1
	BCSC	UNSW	10.5	16.2
	$n^+ pp^+$	Sharp	100	15.8
	$n^+ pp^+$	Telefunken	142	13.5
	commercial	several	100	12.0
FLAT SILICON MODULES				
FZ	BCSC	UNSW	806	16.4
CZ	$n^+ pp^+$	Telefunken	1000	16.8
CZ	commercial	Solarex	5600	12.6
Multicryst.	commercial	Solarex	5600	11.1
GALLIUM ARSENIDE SOLAR CELLS				
GaAs	one junction	NREL	0.2	25.7
GaAs	one junction	Kopin	4	25.1
GaInP/GaAs	two junctions	NREL		27.3

Acronyms: NREL: National Renewable Energies Laboratory, USA. UNSW: University of New South Wales, Australia. PERL: Passivated Emitter and Rear Locally-diffused cell. Back PCC: Back Point-Contact Cell. BCSC: Buried Contact Solar Cell. PESC: Passivated Emitter Solar Cell.

reflectivity lower than 1%. Several methods can be used for the deposition:

- spraying
- screen printing
- spinning
- CVD (chemical vapour deposition)
- vacuum evaporation
- thermal oxidation to grow SiO_2.

A summary of the conversion efficiency achieved by several different cell designs is given in Table 3.2. Gallium arsenide is included for comparison.

3.4.5 Module fabrication

After testing under standard conditions and sorting to match the current and voltage, about 36 cells are interconnected and encapsulated to form a module. The encapsulant materials must satisfy a number of very strict requirements, since the modules should last 20 years or more. A reliable module construction, such as that in Fig. 3.33, has the following components:

1. Front cover: low-iron tempered glass.
2. Encapsulant: transparent, insulating, thermoplastic polymer. The most widely used is EVA (ethylene vinyl acetate).

Fig. 3.33 The module cross section (from P. Hersh and K. Zweibel, *Basic Photovoltaic Principles and Methods*, U.S. Government Printing Office, Washington, DC, 1982. Reprinted with permission)

3. The solar cells and the metal interconnects.
4. Back cover, usually a foil of Tedlar or Mylar.

All these layers are laminated by applying heat and pressure under vacuum. Primers are used to improve adhesion between the layers. The edges are sealed with a neoprene gasket and protected with an aluminium frame. Once finished, the modules are tested and rated according to their power output.

Summary

The crystalline silicon cell technology has been described in steps from sand to final module fabrication. Much of this technology has been adapted from the microelectronics industry but specific methods to produce solar-grade silicon are being developed. In the crystallisation and slicing steps the new techniques are already well established.

Cell- and module-fabrication technology is well developed and reliable. The gap between the efficiencies demonstrated by laboratory prototypes and industrial cells is being narrowed as new cell designs are being transferred from research to production. Modules with efficiencies approaching 18% are becoming available on the market.

3.5 THIN-FILM SOLAR CELLS

3.5.1 Introduction

In comparison with the crystalline silicon cells that were discussed in section 3.4, thin-film technology holds the promise of reducing the module costs through lower material and energy requirements of the manufacturing process. In addition, integrally connected modules are produced directly without the costly individual cell handling and interconnections.

Amongst the large number of possible thin-film cells, a few stand out as candidates for commercial production in the near future. The criteria for commercial viability which we shall now discuss are based mainly on economic considerations and are obviously generalisations. Nevertheless, they suggest technical goals whose achievement would open up very large markets for thin-film solar cells.

Solar radiation has a low power density, and hence the area of PV modules needed to produce significant power output can become quite large. Many costs associated with a photovoltaic generating system increase in proportion to the area of the system. Most module production costs are area-related, and the PV system costs of land, array structures, wiring and transport are all area-related and can dominate the total cost of the system if the efficiency of the modules is low. It is generally considered that module efficiencies over 10% are needed if the system cost is to be held to acceptable levels.

Photovoltaic systems have large initial capital costs but small recurrent costs for operation and maintenance. The cost of delivered energy thus varies inversely as the lifetime of the system. The present PV modules based on silicon wafers exhibit lifetimes of 20–30 years and the target for thin-film cells must be similar.

The annual rate of production of PV modules will reach a few million square metres by the end of the century, and probably a few hundred million square metres in the next 40–50 years. Cells which contain scarce materials could thus be subject to severe constraints due to the availability of these materials at a reasonable price.

The move towards the use of thin-film solar cells is driven by the need to reduce module costs. The cells must therefore be capable of being manufactured in large volumes at low cost. To avoid costly handling, the production process should result in an integrally interconnected module, rather than individual cells requiring separate interconnections.

The final requirement for the solar cell is that it results in no environmental hazard. The hazards must be assessed from the initial mineral extraction, through refining, cell manufacture, PV system operation and final disposal of the modules. The hazards associated with present PV materials are discussed in Chapter 6.

Four types of thin-film cell have emerged to be of likely commercial importance in the next few years (Table 3.3). These are the amorphous silicon cell, most probably in a double-junction structure, thin multicrystalline silicon films on a low-cost substrate, the copper indium diselenide/cadmium sulphide heterojunction cell (or variants of it), and the cadmium telluride/cadmium sulphide heterojunction cell.

Table 3.3 Best results for thin-film solar cells up to 1991 (efficiencies in % under standard AM1.5 illumination)

Material	Commercial products	Best large area (~ 0.1 m^2)	Best R&D small area	Theoretical limit
Amorphous silicon	8–6	10–8.5	13–9	~20
Thin-film silicon	expected 1992	11	16	~25
Copper indium diselenide	expected 1993	11	14	~21
Cadmium telluride	7	10	14	~28

All of these cells have active layers in the thickness range 1–10 μm, and all are manufactured by processes which are capable of large-volume, low-cost production. The materials and the processes were chosen because they had this capability, and the major R&D effort has been to increase the efficiency of the cells and the reproducibility of the product.

Other approaches to thin-film cell manufacture are also being pursued and should be mentioned here. For example, a gallium arsenide thin-film cell can be produced by the CLEFT process. In this process, single-crystal films of gallium arsenide, about 4 μm thick, are grown epitaxially on an (expensive) single-crystal substrate of gallium arsenide. The films are subsequently cleaved from the substrate, which is then available for re-use. It seems at present, however, that these cells will more readily find a market in satellite applications, where cost is less critical than power/mass ratio, rather than in terrestrial use.

3.5.2 Amorphous silicon cells

Amorphous silicon (a-Si) differs from crystalline silicon in that the silicon atoms are not located at very precise distances from each other and the angles between the Si–Si bonds do not have a unique value. The interatomic distances and the bond angles have a range of values with any particular Si–Si distance or bond angle being randomly located within that range.

This randomness in the atomic arrangement has a powerful impact on the electronic properties of the material. It becomes a direct-gap material with an optical energy gap of about 1.75 eV, and its band structure shows a very high density of states within the energy gap, caused largely by incomplete bonding. It was found in 1969 that the incorporation of hydrogen in amorphous silicon could passivate the incomplete bonds and reduce the density of states in the bandgap to such an extent that it became possible to make the material n-type or p-type by the incorporation of phosphorus or boron. Once it became possible to make p-n junctions in a-Si, its high absorption coefficient and ease of manufacture made it an attractive material for solar cells. Both p and n materials have poor transport properties, and simple p-n junction cells have low efficiencies. As we have seen in section 3.4.1, the incorporation of intrinsic a-Si improves the transport properties, and the development of p-i-n junction cells led to rapid improvements in performance. The i-layer is not in fact completely intrinsic, but is slightly n-type. It is therefore preferable to have light enter the cell through a very thin p-type layer so that the region of maximum photogeneration in the i-layer is also the region of highest electric field, between the heavily doped p^+-layer and the very lightly n-doped 'i'-layer.

The optimum structure of the a-Si cell is the p^+-i-n, which was already discussed in section 3.3, has a transparent conducting contact to the p^+-layer and an ohmic contact to the n-layer, as shown in Fig. 3.34. Unfortunately, an optically transparent highly conductive p-type material has not yet been discovered. The transparent conducting material used is usually tin oxide,

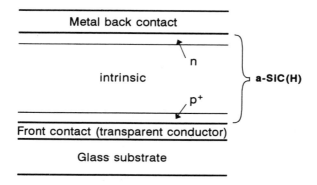

Fig. 3.34 The structure of amorphous silicon p-i-n solar cell

which is n-type. However, the highly doped TCO (transparent conducting oxide) layers form a tunnel junction with the p⁺ a-Si layer, and the resistance to current flow is very small.

Having determined the optimum structure, the manufacturing process sequence can be defined (Fig. 3.35). Starting with glass as a low-cost transparent, weatherproof substrate, a layer of highly conductive optically transparent tin oxide is deposited, then a highly doped p-layer of a-Si sufficiently thin to absorb little light, then the undoped, but slightly n-type intrinsic layer of a-Si, then a thin conductive n-layer of a-Si and finally a metallic contact layer. This contact layer should form an ohmic contact to the n-layer of a-Si and ideally should also be highly reflective so that any light not absorbed in passing through the cell is reflected back for a second pass through the i-layer. Silver would be the best material except for its cost and aluminium or aluminium alloys are usually used. The original production process which produced the a-Si/H_2 alloy, needed for semiconductor-grade amorphous silicon, was the plasma decomposition of silane (SiH_4). This is still the best process, producing the most efficient cells, although other processes such as sputtering have been studied in some detail, and elements such as fluorine have also been proposed as passivators of the dangling bonds.

As the quality of the a-Si material improved through the 1970s, it became apparent that the action of light absorption by the i-layer created additional defects, increasing the density of trapping and scattering states and reducing the efficiency of the cells. This effect—named the *Staebler–Wronski effect* after its discoverers—is dependent on the total number of photons absorbed. It therefore depends on the intensity of the light to which the cell is exposed, the duration of the exposure and on the thickness of the i-layer. Exposure to room lighting, as with solar calculators, etc., has only a small effect, but bright sunlight reduces the efficiency considerably over times-cales of months. This instability has serious consequences for the commer-

61

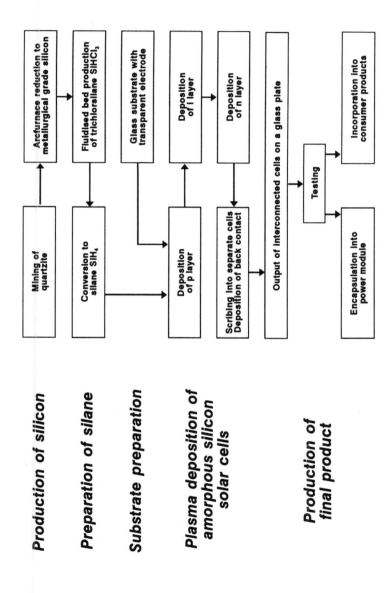

Fig. 3.35 Production of amorphous silicon PV modules

cial viability of a-Si as a power producer and it now seems unlikely that the single-junction a-Si cell will be used in such applications.

Since the Staebler–Wronski effect depends on the thickness of the i-layer, it can be alleviated by using multiple-junction structures, in which the absorption of light is split equally between two or three separate i-layers. The simplest of such structures is just a p-i-n/p-i-n stack of a-Si(H), shown in Fig. 3.36, in which the overall thickness of the cell is maintained at about 1 μm, with the front i-layer about 0.3 μm and the back i-layer about 0.7 μm, to produce equal currents from each half. Cells with these structures have been produced commercially and found to be significantly more stable than the single-junction device. It is not yet clear whether the stability is sufficient to allow these cells to be commercially viable as power generators.

Amorphous silicon forms good alloys with both carbon and germanium. The bandgap of the alloy a-SiC(H) is increased in proportion to the fraction of carbon whilst that of a-SiGe(H) is reduced in proportion to the fraction of germanium. Both alloys can be produced by the same plasma deposition process which deposits a-Si(H) with just the feed silane being replaced by a mixture of silane/methane (SiH_4/CH_4) or silane/germane (SiH_4/GeH_4). This gives the possibility of many multijunction structures, for instance a p-i-n front cell of a-SiC(H), a middle p-i-n cell of a-Si(H) and a back p-i-n cell of a-SiGe(H) shown in Fig. 3.37. Such a cell has been produced with an efficiency of 13.3%, compared to the best single-junction efficiency of 12%.

The principal technical difficulty with these cells is the formation of the tunnel junction between the n-layer of the upper cell and the p-layer of

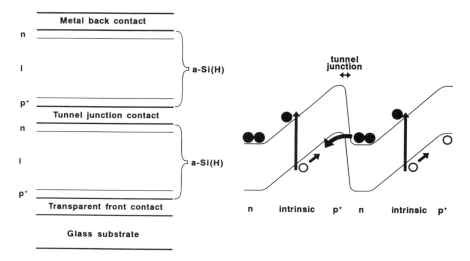

Fig. 3.36 The structure and band diagram of the amorphous silicon p-i-n/p-i-n solar cell

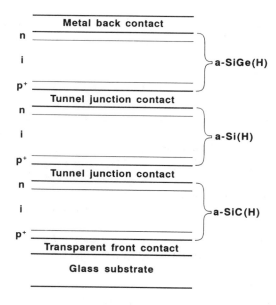

Fig. 3.37 Triple-stack amorphous silicon solar cell

the lower cell. Good results can be achieved in the laboratory on small-area cells, but it is a major challenge to achieve similar results reliably and repeatably in commercial production on large-area modules.

3.5.3 Thin polycrystalline silicon on low-cost substrates

There is good theoretical justification for using thin base layers in silicon solar cells because the dark saturation current I_0 decreases with decreasing base-layer thickness, leading to higher open-circuit voltage values for thin cells. The problem, at first sight, is that complete optical absorption requires thicknesses of a few hundred micrometres. As we have seen in section 3.3, it is nevertheless possible to maintain optical path lengths of hundreds of micrometres with silicon layers of tens of micrometres thick by using light-trapping techniques.

Light incident on a flat planar cell will enter through an antireflection (AR) coating and an exponentially decreasing proportion will travel through the cell to the back contact. Some light incident on the back contact will be reflected back through the cell for a second pass in which it will be either absorbed or leave the cell via the AR coating. If the top and back surfaces are textured, rather than being flat, the light entering through the AR coating will be refracted off-normal, and light reaching the back will be scattered, as will light reaching the top surface after reflection from the back. In this way, the optical path length of a typical light ray normally incident on the cell can be up to 20 times the cell thickness. Complete optical absorption can then be assured even for cell thicknesses of only a few tens of micrometres.

There are other benefits to such thin silicon cells. The diffusion length need be only 50–80 µm, so lower quality material can be used, and higher doping levels can be tolerated, giving higher open-circuit voltage. Because the optical absorption, and hence photogeneration, is taking place within or close to the junction depletion region, carrier collection efficiencies are high, giving high short-circuit currents. The reduced constraints on material quality mean that film deposition of silicon layers on low-cost substrates can be considered without sacrificing much in the performance of the cells. This is therefore a very promising route for the production of efficient cells at low cost.

There are a number of key areas to the successful implementation of the concept. One needs a low-cost substrate which has a good thermal expansion match to silicon, to avoid strain in the silicon film after processing, and one needs a silicon film-deposition process which gives high yields of suitable materials and potentially high throughput at acceptable capital cost.

The growth of silicon films from saturated solutions of silicon in molten tin by *liquid phase epitaxy* (LPE), already mentioned in section 3.4, has been found to meet the requirements for a commercial production process. The films as-grown have a columnar polycrystalline structure and post-processing is needed to increase polycrystallite size and passivate the grain boundaries. After this treatment, the p-type silicon film is processed by conventional diffusion techniques to give an n-p junction, then AR coated and top contacted in the usual way. The light trapping arises from the topology of the substrate surface, and the optical properties of the back contact metallisation. Efficiencies up to 16% have been observed in small-area cells and about 12% in commercial 10 cm × 10 cm cells.

The process as it stands produces the equivalent of a low-cost wafer silicon cell which must be interconnected at the subsequent module production stage. It is possible to envisage a module-sized ceramic substrate with cells deposited simultaneously on the substrate with a subsequent metallisation/interconnection step for the whole module. The key to the realisation of this advance is the availability of ceramic substrates of suitable size. If these should become available, the other technical problems of scaling up the LPE process are challenging but solvable, and thin-silicon technology could be an important route to a module price of $1 per peak watt.

3.5.4 Copper indium diselenide cells

Copper indium diselenide (CIS) is a semiconducting material which can be either n- or p-type and has a direct optical absorption with the highest absorption coefficient measured to date. The electronic properties of CIS depend critically on the copper/indium ratio and good control of the stoichiometry is essential for efficient devices.

It is possible to make p-n homojunctions of CIS, but these are neither stable nor efficient, and the best devices to date are heterojunctions with cadmium sulphide (CdS). There is quite a good match between the lattice constants and electron affinities of CIS and CdS so the recombination at their interface is not excessive. CdS can only be grown as n-type material, so the CIS must be p-type. It is fairly easy to make ohmic contact to CdS, particularly when it is heavily doped, but contacts to the p-type CIS are more difficult. Gold makes a good ohmic contact but is too expensive for commercial use. Molybdenum is the metal most commonly used, and makes reliable contacts to heavily doped p-type CIS. For efficient devices, the CIS needs to be fairly lightly doped near the junction, so that the depletion region extends throughout the volume where photogeneration takes place. This ensures field-aided collection of the photogenerated carriers and hence high short-circuit currents.

CIS has a bandgap of 1 eV and grows as a columnar polycrystalline film. The CdS also grows, to some extent topotaxially, as a columnar polycrystalline film on top of the CIS. The open-circuit voltage of the cell is thus very dependent on the recombination currents near the junction, due to the grain boundaries and the interface. Good quality CIS has grain sizes of a few micrometres and diffusion lengths of around 2 μm, so passivation of the grain boundaries is essential to reduce the recombination currents. An oxidation treatment after deposition, for instance heating in air, can give significant improvements in performance, due to grain boundary passivation, and clearly a deposition method which gives large grains with minimum grain boundary states is advantageous.

CdS has a bandgap of 2.4 eV so it will absorb strongly all incident radiation in the green/blue end of the spectrum. The optimum devices use a very thin (0.03 μm) layer of CdS with a wide-gap window layer of highly conductive material. Zinc oxide has been found to be a very suitable material for the window layer, so the structure of the CIS cell is that shown in Fig. 3.38.

The production sequence of CIS modules is shown in Fig. 3.39. In the first ten years of R&D on CIS cells, the deposition method used by all workers was vacuum evaporation of the elements. It was quickly discovered that maintenance of stoichiometry depended on providing the correct copper/indium ratio and supplying an excess of selenium vapour pressure above the growing film surface. A later development with better potential for commercial production was to deposit metallic indium and copper on to the molybdenum contact metal and then selenise the indium/copper layers by passing hydrogen selenide gas over the surface at about 400°C. This selenisation process gives high-quality material and can produce uniform

Fig. 3.38 Structure of CIS solar cell

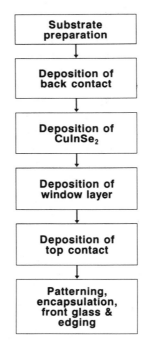

Fig. 3.39 Production of CIS modules

films over large areas, suitable for module production. The incorporation of selenium into the indium/copper layer causes a large increase in volume, and can result in large stresses in the CIS film, even to the extent of delamination of the film from the substrate. These problems have been addressed by careful control of the details of substrate preparation, molybdenum deposition, sputtering of the copper and indium, and control of the selenisation process. The addition of some gallium at the Mo/CuIn interface is also found to improve the adhesion quite markedly.

The incorporation of gallium into CIS has a number of benefits, apart from improved adhesion of the film to the molybdenum. Gallium substitutes for indium to give a true alloy—copper indium/gallium diselenide (CIGS)—with a bandgap which increases with increasing gallium fraction. As the bandgap increases, the open-circuit voltage increases, giving better fill factors and fewer cells per module. For CIS, the voltage at maximum power is usually only about 300–350 mV, which is inconveniently low. CIGS with 10–20% gallium can raise this to values of 450–500 mV, about the same as crystalline silicon cells, and, with fewer cells to be integrally interconnected in module production, the processing is more reliable and less costly. CIGS with a bandgap of greater than 1.1 eV also avoids problems with the free-carrier absorption in the zinc oxide window. The conductance of zinc oxide is due to free electrons and the long-wavelength absorption due to these free carriers begins at about 1.1 µm (1.1 eV). It is possible to reduce the free carrier absorption by improving the quality of the ZnO film, but it cannot be eliminated. However, with CIGS, the problem no longer exists since photons of energy of 1.1 eV or less are not absorbed by the CIGS, and the current lost by the increased bandgap is more than compensated for by the increased voltage and fill factor.

A considerable effort has been devoted to scaling up the CIS technology to module size, mainly by Siemens Solar Industries (formerly ARCO Solar). Square modules of area 0.1 m² have been produced for some time and subjected to rigorous indoor and outdoor testing. SSI can now produce modules four times this size with efficiencies over 9%, which appear to be stable over periods of some years.

3.5.5 Cadmium telluride cells

It has been known since the 1950s that cadmium telluride has the ideal bandgap for a solar absorber material and the research on these cells can be traced back to that time. These efforts, however, were small and uncoordinated and severely hampered by the low level of materials technology for the II/VI materials. As the II/VI materials technologies improved over the years, it became clear that cadmium telluride performance was limited by a high level of defect states close to the middle of the bandgap, forming very efficient recombination sites. The achievement of the late 1970s was to develop post-deposition treatments which greatly reduced the density of these recombination states, and the achievement of the 1980s has been to capitalise on this earlier work to develop efficient cells and efficient stable modules.

There are a number of low-cost techniques which can be used to deposit cadmium telluride and they can all, after a post-deposition treatment, yield high-quality material and efficient cells. The choice of technology for commercial production can then be made on the basis of its fitness for the expected markets, environmental considerations and the availability of skills, capital, etc., within the production company.

There are at present three companies close to production of CdTe devices: Matsushita in Japan, Photon Energy in the USA and BP Solar in Europe. Matsushita have developed screen printing technologies for all of the films in the cell, back and front contacts, cadmium telluride and cadmium sulphide. Screen printing of solar cells offers a technology with low-cost production at quite low production rates. The rate-limiting step is the heat treatment of the printed inks after deposition, which requires fairly high temperatures (around 500°C) for periods of an hour or more. The key to the success of the screen printing process seems to be the use of elemental cadmium and tellurium in the inks, rather than CdTe powder. The Cd and Te react during the sintering process to produce CdTe and the heat of reaction is high enough to promote the growth of high-quality layers. These screen-printed cells have been sold for some years for use in solar calculators, but there are problems in scaling up to power modules because of the large degree of contact obscuration (40%) in this technology, although modules with an aperture efficiency of 6% have been produced and tested in outdoor conditions for some years, and are commercially available in limited quantities.

Fig. 3.40 Structure of CdTe solar cell

Photon Energy is a small company with many years of expertise in the deposition of II/VI materials by spraying and other wet chemical methods. The layers in their integrally interconnected modules are deposited by low-cost processes and Photon Energy estimates production costs at low volume to be about $70/m^2, falling to $50/m^2 at volumes of around 100 000 m^2 per annum. CdTe/CdS cells suffer from the absorption of green/blue light by thick CdS just as do CIS cells and Photon Energy have developed a novel structure with thin CdS, shown in Fig. 3.40, which has resulted in a small-area cell of 12.3% efficiency. Modules of around 8% efficiency have been subject to independent stability tests for up to 300 days outdoors and have been found to show no degradation in performance over that time.

BP Solar has developed the electroplating technology originally devised by Monosolar, a company which they purchased in 1984. Electroplating is a particularly favourable technology for CdTe cells because of its very high utilisation of material and low capital cost of the equipment relative to vacuum-deposition processes. It is relatively easy to scale up the process to produce module-sized layers and it should be capable of low-cost production at low volumes. BP Solar have produced modules of area about 0.1 m^2 with efficiencies over to 10% and 1 cm^2 cells with efficiencies of around 13%. Outdoor stability tests have been performed in Spain and application development tests elsewhere with excellent results.

There is a variety of vacuum deposition techniques which have produced good results in the research laboratory on small-area cells, but only one of these—closed space sublimation—is being considered for potential commercial development. It is envisaged that ultimately thin-film solar cells will be produced in a facility at the end of a float glass production line with a throughput of around 2 million m^2 per annum. The temperature of glass leaving the float glass production line would be ideal for the closed space sublimation process and, with these production volumes, the cost per unit area would be small.

3.5.6. Integrally interconnected modules

One of the advantages of most thin-film solar cells is that the full module can be deposited with the cell interconnections made during deposition. The technique is illustrated in Fig. 3.41. The tin oxide coating on the glass is scribed, either mechanically or by laser, into a set of strips 1–2 cm wide running the full length of the module. The a-Si p-i-n junction or multijunctions are next deposited over the full area, and then scribed, with the scribe lines slightly offset from those of the tin oxide. The metal back contact is then deposited over the whole area, and then scribed with a further offset. Each individual cell is in the form of a long narrow strip, which reduces series resistance, and all cells are connected in series, with the outer metal

Glass substrate

Transparent
front contact

Metal back
contact

Solar cell
material

Fig. 3.41 Integrally interconnected module

strip being the negative terminal and the outer tin oxide strip the positive terminal for the whole module.

This technique reduces the handling required to a very considerable extent and reduces the manufacturing costs of thin-film modules quite significantly.

Summary

We have discussed the operation and manufacture of four types of thin-film cells which are in or near commercial production. All these cells have the potential for large-scale production at low unit cost. The reasons for the limited stability of amorphous silicon cells have been analysed, and a variety of structures under development which may overcome these problems have been discussed. We have shown that, with currently available technology and theoretical understanding, the thickness of polycrystalline silicon cells can be greatly reduced, leading to a promising new route for the production of efficient cells at a lower cost. We have also considered the structure and manufacture of two compound thin-film cells, CIS and CdTe. Finally, we have examined an important part of the thin-film photovoltaic technology: the manufacture of integrally interconnected modules.

SUMMARY OF THE CHAPTER

Solar cells convert sunlight into electricity by exploiting the photovoltaic effect. The maximum conversion efficiencies observed in laboratory devices have exceeded 30% but the typical efficiencies for production cells are 10–15%.

The physical principles of photovoltaic energy conversion have been discussed, starting from electronic properties of semiconductors. A variety of structures have been examined with a view of how power losses which accompany their operation can be reduced.

Crystalline silicon solar cells represent the largest part of the market. The technology has been reviewed in detail, starting from crystalline silicon production through wafer preparation and cell manufacture to the final module assembly. The traditional single-crystal technology, as well as the newer semicrystalline and ribbon techniques, have been examined. We have also taken a brief look at some devices in the research stage which are likely to shape the technology in the years to come.

Thin-film solar cells hold the promise of inexpensive technology with acceptable conversion efficiencies. The requirements on a thin-film cell have been considered. Four types of thin-film cells that are to be of likely commercial importance in the next few years have been examined in detail: the amorphous silicon cell, thin polycrystalline silicon cell grown on a low-cost substrate, the copper indium diselenide cell and the cadmium telluride cell.

BIBLIOGRAPHY AND REFERENCES

GREEN, M. A., *Solar Cells*, Prentice Hall, Englewood Cliffs, NJ, 1982.
HERSH, P. and ZWEIBEL, K., *Basic Photovoltaic Principles and Methods*, U.S. Government Printing Office, Washington, DC, SERI/SP-290-1448, 1982.
PULFREY, D. L., *Photovoltaic Power Generation*, Van Nostrand Rheinhold, New York, 1978.
VAN OVERSTRAETEN, R. and MERTENS, R., *Physics, Technology and Use of Photovoltaics*, Adam Hilger, Bristol, 1986.

SELF-ASSESSMENT QUESTIONS

PART A. True or false?

1 Majority carriers are the principal mobile charges created by light in semiconductors.

2   The conduction band is so-called because its structure determines the electrical properties of a semiconductor.

3   Boron contamination must be carefully avoided in silicon solar cell manufacture.

4   Cells made from low-bandgap semiconductors usually produce large current.

5   A solar cell can be pictured as a current generator connected in series with a diode.

6   The size of contacts in the point-contact cell aims to reduce the obscuration of the cell.

7   The main disadvantage of amorphous silicon for solar cell manufacture is its low light absorption.

8   The voltage output of a solar cell is directly proportional to the irradiance.

9   Recombination represents the principal voltage loss mechanism in good solar cells.

10   Series resistance reduces mainly the fill factor of a solar cell.

11   The front cell in a tandem must be made of high-bandgap semiconductor.

12   Light trapping is a means of increasing poor light absorption properties of most thin films.

13   Pure silicon is expensive because silicon ore is rare.

14   The slicing step is what limits the throughput of all ingot growth methods.

15   The major drawback of ribbon growth methods is their throughput.

16   The textured surface of a cell is rough as a result of the sawing step.

17   The reflectivity of bare silicon is high but decreases to acceptable levels after encapsulating the cells beneath glass.

18   The p-i-n structure of amorphous silicon cells is used for the beneficial effect of electric field in the region where carriers are generated.

19   CIS cells owe their importance to the fact that the thin film has good crystalline properties and no treatment of the grain boundaries is necessary.

20   The CdS layer in compound thin-film cells generates much of the current produced by the cell.

PART B.

1   What are intrinsic, n-type and p-type semiconductors?

2   Define the efficiency of a solar cell. Under what conditions it is usually measured?

3   Write down the I–V characteristic of an ideal solar cell. How does it differ from the I–V characteristic of a diode?

4   Using Table 3.1, would you expect the open-circuit voltage of a crystalline silicon cell to be smaller or larger than for an amorphous silicon cell?

5   What are the principal current losses in a silicon solar cell?

6   What measures can be taken to minimise reflection from the top surface of the cell?

7   What is the highest efficiency attained by a silicon cell under standard test conditions?

8   List the five criteria which define the ideal thin-film cell.

9   Why does an amorphous silicon cell degrade under sunlight?

10   How can amorphous cells be made more efficient and more resistant to photodegradation?

11   How is light absorbed efficiently in thin polycrystalline silicon cells?

12   How is the performance of CIS cells improved by alloying with gallium?

13   What methods are used to deposit CdTe for solar cells?

14   What are the advantages of producing integrally interconnected modules?

Answers

Part A

1, False; 2, False; 3, False; 4, True; 5, False; 6, False; 7, False; 8, False; 9, True; 10, True; 11, True; 12, False; 13, False; 14, True; 15, False; 16, False; 18, True; 19, False; 20, False.

Part B

1   Intrinsic semiconductors are undoped; n-type and p-type contain donor and acceptor impurities, respectively. The electric current is carried predominantly by electrons in n-type, and by holes in p-type.

2   The efficiency of a solar cell in percent is the power output in milliwatts at the maximum power point under AM1.5 (100 mW/cm^2) illumination, at 25°C.

3   $I = I_\ell - I_o [\exp(qV/kT) - 1]$. The diode characteristic does not contain I_ℓ.

4   Smaller since the bandgap of crystalline silicon is smaller.

5   Top-contact shading, top-surface reflection, collection losses by recombination, and incomplete absorption of light.

6   Antireflection coating (preferably in several layers) and/or surface texturing.

7   23.3%

8   (a) Reasonable efficiency (over 10%). (b) Long lifetime (20 years or more). (c) Must not contain scarce materials. (d) Should produce integrally interconnected modules. (e) No environmental damage or hazard.

9 Light creates additional recombination or scattering centres (the Staebler–Wronski effect), thus reducing the cell output, particularly under bright sunlight.

10 Using structures formed by several thin layers based on a-S(H) or its alloys with carbon or germanium.

11 By exploiting the light-trapping effect, with reflecting back surface and top surface textured.

12 The alloy gives higher open-circuit voltage, better adhesion to the molybdenum contact, and lesser effect of light absorption in the window ZnO layer.

13 (a) Screen printing. (b) Spraying or other wet chemical methods. (c) Electroplating.

14 The module can be manufactured as a part of the cell production process, reducing the individual cell handling and interconnection and thus the total cost.

4

Photovoltaic Engineering

AIM

The aim of this chapter is to survey the structure, operation and design of photovoltaic systems

OBJECTIVES

On completion of this chapter, you should be able to:

1. Understand the structure of a photovoltaic system, its subsystems, and the concept of sizing.
2. Analyse the operation of the photovoltaic generator in terms of modules and their electrical characteristics.
3. Examine the battery operation in photovoltaic systems.
4. Evaluate the role of power conditioning and control elements.
5. Size a simple system using the radiation power supply and load power requirements.

NOTATION AND UNITS

Symbol		SI unit	Other unit
E_L	Daily load energy requirement	J	Wh
FF	Fill factor		
G	Solar irradiance	W/m^2	
I_L	Average load current	A	

Edited from manuscripts of L. Castañer (Photovoltaic engineering), E. Lorenzo (Photovoltaic generator, Photovoltaic systems) and M. A. Egido (Sizing). Additional information by A. Sorokin and F. C. Treble.

(Contd.)

Symbol		SI unit	Other unit
I_m	Module current at maximum power point	A	
I_p	PV generator current at maximum power point under standard conditions	A	
I_{sc}	Short-circuit current	A	
k	Boltzmann constant	J/K	eV/K
LLP	Loss-of-load probability		
NOCT	Normal operating cell temperature	K	C
N_s	Number of series-connected modules		
N_p	Number of parallel strings		
n_c	Number of cells in a module		
P_{max}	Power at maximum power point	W	
PSH	Peak Solar Hours		h
Q_{yd}	Yearly charge deficit	C	Ah
Q_{los}	Charge deficit to compensate for lack-of-sunshine days	C	Ah
Q_B	Nominal battery capacity	C	Ah
R	Load resistance	Ω	
R_s	Series resistance	Ω	
SF	Array oversize (safety) factor		
T_a	Ambient temperature	K	C
T_c	Cell operating temperature	K	C
t_i	Duration of operation of appliance i	s	h
V_{DC}	DC bus-bar voltage	V	
V_{oc}	Open-circuit voltage	V	
V_R	Voltage at resistive load	V	
V_m	Voltage at maximum power	V	
W_i	Nominal power rating of appliance i	W	
ΔE	Yearly energy deficit	J	Wh
Φ	Lowest permitted state of charge of the battery		

Unit conversion factors

$$\text{To convert} \begin{Bmatrix} \text{charge} \\ \text{energy} \\ \text{temperature} \end{Bmatrix} \text{from} \begin{Bmatrix} \text{Ah} \\ \text{Wh} \\ \text{°C} \end{Bmatrix} \text{to} \begin{Bmatrix} \text{C} \\ \text{J} \\ \text{K} \end{Bmatrix} \begin{matrix} \text{divide by 3600} \\ \text{divide by 3600} \\ \text{add 273.16} \end{matrix}$$

Values of physical constants

$$k = 1.38 \times 10^{-23} \text{ J/K} = 86.3 \times 10^{-3} \text{ eV/K}$$

4.1 INTRODUCTION

Solar cells are not the only components of a PV system. Many other parts are usually required to provide a satisfactory electricity supply.

Many PV systems contain a provision for energy storage to supply electricity at night and during periods of inclement weather.

Solar cells generate direct current. Since most available appliances work with alternating current, some form of power conditioning is usually required.

Other power conditioning or control elements are needed to interface the different parts of the system, and to allow for the variable nature of the converted solar energy. Some elements also reflect the desirability to monitor the performance.

All these components have to be properly interconnected, sized and specified for PV operation.

In this unit, you will be given an overall view of the photovoltaic system, and how these problems are addressed in its design.

In section 4.2, we shall examine the structure of a system, and how the various parts, or subsystems, work together to provide the desired power output.

Each subsystem will then be analysed in more detail in sections 4.3–4.5. Their construction and components will be reviewed, and the main characteristics presented to you. You will be shown how solar cells, which were the subject of Chapter 3, can be integrated into a *photovoltaic generator*. We shall highlight the photovoltaic aspects of the storage and power conditioning subsystems which are also standard components of other energy storage and control systems.

The size of the system and, indeed, of the PV generator and storage subsystems, depend on the geographical location and on the application for which the system is intended. Solar energy can be a very reliable power source in isolated locations, with the minimum attention and maintenance. The reliability aspect of a PV system is therefore very important. As we shall see in this chapter, many decisions that go into the design of a system represent a careful consideration between the quest for maximum power production, and reliability. In particular, *sizing* of a photovoltaic system is an important part of its design, and you will be shown how it is done in section 4.6.

4.2 STRUCTURE OF A PHOTOVOLTAIC SYSTEM

The photovoltaic system consists of a number of parts or subsystems (Fig. 4.1):

(a) The photovoltaic generator with mechanical support and, possibly, a sun-tracking system.
(b) Batteries (storage subsystem).
(c) Power conditioning and control equipment, including provision for measurement and monitoring.
(d) Back-up generator.

Fig. 4.1 The photovoltaic system. Usually, of course, not all the subsystems will be included

The choice of how and which of these components are integrated into the system is governed by various considerations.

There are two main categories of systems, *grid-connected* and *stand-alone*. The simplest form of the latter consists simply of a photovoltaic generator alone which supplies DC power to a load whenever there is adequate illumination. This type of system is common in pumping applications. In other instances, the system will usually contain a provision for energy storage by batteries. Some form of power conditioning is then frequently also included, as is the case when AC current is required at the output from the system. In some situations, the system contains a back-up generator.

Grid connected systems can be subdivided into those in which the grid merely acts as an auxiliary supply (*grid back-up*) and those in which it may also receive excess power from the PV generator (*grid interactive*)*. In PV power stations, all the generated power is fed into the grid.

Summary

We have reviewed the basic types of photovoltaic systems, and introduced the principal subsystems. Distinction has been made between grid-connected and stand-alone systems.

---

* or synchronously connected, in the language of electrical engineers.

4.3 THE PHOTOVOLTAIC GENERATOR

4.3.1 Introduction

The heart of the system is the photovoltaic generator. It consists of *photovoltaic modules* which are interconnected to form a DC power-producing unit. The physical assembly of modules with supports is usually called an *array* (Fig. 4.2).

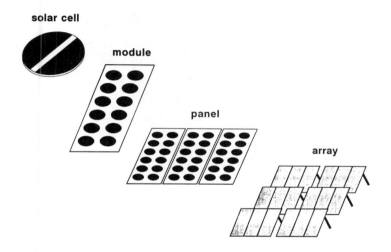

Fig. 4.2 The photovoltaic hierarchy

The module construction was introduced in Chapter 3. In section 4.3.2 of this chapter, we examine the electrical characteristics of a module in detail and show how to determine the module output during practical operation.

Module interconnection in a photovoltaic generator will be discussed in section 4.3.3 where we describe the characteristics of a photovoltaic generator under conditions that are encountered in practical operation.

In section 4.3.4, we discuss the two principal options for panel orientation: fixed inclination, or tracking.

4.3.2 Photovoltaic modules

The photovoltaic module was already encountered in chapter 3. As the name suggests, the module represents the basic construction unit of a

Fig. 4.3 The photovoltaic module based on crystalline silicon solar cells

photovoltaic generator. The structure of a module based on crystalline or semi-crystalline silicon cells (discussed in section 3.4.5) is shown in Fig. 4.3. It is this type of module that is most commonly used in practice, and we shall consider its electrical characteristics in more detail.

Most frequently, the cells in a module are interconnected in series. The reason comes from the electrical characteristics of an individual solar cell. As we have seen in Chapter 3, a typical 4-inch diameter crystalline silicon solar cell, or a 10 cm × 10 cm multicrystalline cell, will provide between 1 and 1.5 watts under standard conditions, depending on the cell efficiency. This power is usually supplied at a voltage 0.5 to 0.6 V. Since there are very few appliances that work at this voltage, the immediate solution is to connect the solar cells in series.

The number of cells in a module is governed by the voltage of the module. The nominal operating voltage of the system usually has to be matched to the nominal voltage of the storage subsystem. Most of photovoltaic module manufacturers therefore have standard configurations which can work with 12 volt batteries. Allowing for some overvoltage to charge the battery and to compensate for lower output under less-than-perfect conditions, it is found that a set of 33 to 36 solar cells in series usually ensures reliable operation.

The power of silicon modules thus usually falls between 40 and 60 W. The module parameters are specified by the manufacturer under the following standard conditions:

$$
\begin{array}{ll}
\text{Irradiance} & 1 \text{ kW/m}^2 \\
\text{Spectral distribution} & \text{AM1.5} \\
\text{Cell Temperature} & 25°\text{C}
\end{array} \tag{4.1}
$$

Indeed, they are the same conditions as are used to characterise solar cells. The nominal output is usually called the *peak power* of a module, and expressed in *peak watts*, W.

The three most important electrical characteristics of a module are the short-circuit current, open-circuit voltage and the maximum power point as functions of the temperature and irradiance. These characteristics resemble the *I–V* characteristic of a solar cell (Fig. 4.4) but some specific features need to be highlighted.

Temperature is an important parameter of a PV system operation. As we have seen in Chapter 3, the temperature coefficient for the open-circuit voltage is approximately equal to − 2.3 mV/°C for an individual cell. The voltage coefficient of a module is therefore negative and very large

Fig. 4.4 The temperature and irradiance dependence of the module *I–V* characteristic

since 33 to 36 cells are connected in series. The current coefficient, on the other hand, is positive and small—about + 6 $\mu A/°C$ for a square centimetre of the module area. Accordingly, only the voltage variation with temperature is allowed for in practical calculations, and for an individual module consisting of n_c cells connected in series is set equal to

$$d V_{oc}/d T = - 2.3 \times n_c \text{ mV}/°C \qquad (4.2)$$

It is important to note that the voltage is determined by the operating temperature of the cells which differs from the ambient temperature (see equation (4.5) below).

As for a single cell, the short-circuit current I_{sc} of a module is proportional to the *irradiance*, and will therefore vary during the day in the same manner. Since the voltage is a logarithmic function of the current, it will also depend logarithmically on the irradiance. During the day, the voltage will therefore vary less than the current. In the design of the PV generator, it is customary to neglect the voltage variation, and to set the short-circuit current proportional to the irradiance:

$$I_{sc}(G) = I_{sc} \text{ (at 1 kW/m}^2\text{)} \times G \text{(in kW/m}^2\text{)} \qquad (4.3)$$

The operation of the module should lie as close as possible to the *maximum power point*. It is a significant feature of the module characteristic that the voltage of the maximum power point, V_m, is roughly independent of irradiance. The average value of this voltage during the day can be estimated as 80% of the open-circuit voltage under standard irradiance conditions. This property is useful for the design of the power conditioning equipment.

The characterisation of the PV module is completed by measuring the *Normal Operating Cell Temperature* (NOCT) defined as the cell temperature when the module operates under the following conditions at open circuit:

Irradiance	0.8 kWm^2	
Spectral distribution	AM1.5	(4.4)
Ambient temperature	20°C	
Wind speed	> 1 m/s	

NOCT (usually between 42°C and 46°C) is then used to determine the solar cell temperature T_c during module operation. It is usually assumed that the difference between T_c and the ambient temperature T_a depends linearly on the irradiance G in the following manner:

$$T_c - T_a = \frac{\text{NOCT} - 20}{0.8} G(\text{kW/m}^2). \qquad (4.5)$$

Fig. 4.5 summarises the calculation of module parameters under operating conditions. The short-circuit current is determined from equation (4.3). The

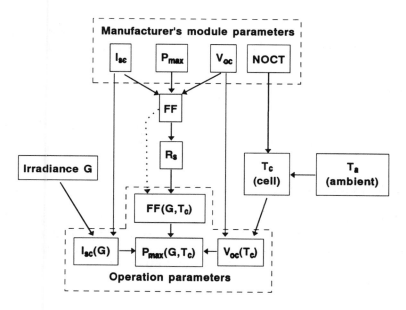

Fig. 4.5 Calculation of the module operation parameters

cell operating temperature is found using equation (4.5) and the resulting value is then used to obtain the open-circuit voltage by using (4.2). The temperature and irradiance dependence of the fill factor is complicated, but can be determined by calculating first the series resistance which is independent of the module conditions. In a simplified calculation, the fill factor can be assumed constant. These parameters then yield the maximum power of the module under operating conditions.

4.3.2.1 Worked example

Determine the parameters of a module formed by 34 solar cells in series, under the operating conditions $G = 700$ W/m, and $T_a = 34°C$. The manufacturer's values under standard conditions are: $I_{sc} = 3$ A; $V_{oc} = 20.4$ V; $P_{max} = 45.9$ W; NOCT $= 43°C$

Solution

1. Short-circuit current
   $I_{sc} = (700$ W/m$^2) = 3 \times 0.7$ (kW/m^2) $= 2.1$ A.

2. Solar cell temperature
   $T_c = 34 + 0.7 \times (43 - 20)/0.8 = 54.12°C$.

3. Open-circuit voltage
   $V_{oc}(54.12°C) = 20.4 - 0.0023 \times 34 \times (54.12 - 25) = 18.1$ V

4. We shall now determine the maximum power point using the simplifying assumption that the fill factor is independent of the temperature and the irradiance:

$$FF = 45.9/(3 \times 20.4) = 0.75$$
$$P_{max}(G, T_c) = 2.1 \times 18.1 \times 0.75 = 28.5 \text{ W}.$$

Thus, noting the manufacturer's value of P_{max} we see that the module will operate at about 62% of its nominal rating.

4.3.3 Interconnection of PV modules

A schematic diagram of a PV generator consisting of several modules is shown in Fig. 4.6. In addition to photovoltaic modules, the generator contains by-pass and blocking diodes. As we shall see below and in section 4.4, these diodes protect the modules and prevent the generator acting as a load in the dark.

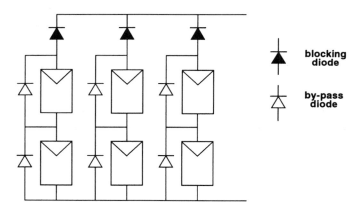

Fig. 4.6 The photovoltaic generator

The modules are connected in series to form strings, where the number of modules N_s is determined by the selected DC bus voltage, and the number of parallel strings N_p is given by the current required from the generator. For the generator in Fig. 4.6, $N_s = 2$ and $N_p = 3$. The output voltage would therefore be twice the module voltage, and the current three times the module current. We shall discuss these points further in section 4.6.

This analysis assumes that all the modules are identical. In practice, the modules (and cells) are not identical, and their parameters exhibit a certain degree of variability for two principal reasons:

- The solar cells and modules vary in quality as a result of the manufacturing process. In general, the current produced by commercial modules suffers a higher degree of dispersion than the voltage.
- Different operating conditions may exist in different parts of the PV array. For example, one must allow for different cleanness of different parts of the PV generator, or some modules may be obscured by a cloud which is covering only a part of the array.

This variability of component parameters has two important effects. Firstly, the output power of the generator is less than the sum of the values corresponding to all the constituent modules. This gives rise to *mismatch losses*. These losses can be minimised by sorting the modules prior to interconnection, and forming series strings from modules with similar values of short-circuit current.

Secondly, there is a potential for overheating the 'poorest' cell of a series string. In some circumstances, a cell can operate as 'load' for other cells acting as 'generators'. Consequently, this cell dissipates energy and its temperature increases. If the cell temperature rises above a certain limit (85–100°C) the encapsulating materials can be damaged, and this will degrade the performance of the entire module. This is called the *hot spot* formation.

This effect is illustrated in Fig. 4.7 which shows a cell in a string which does not produce current—this can happen, for example, when the cell is shaded. We learned in Chapter 3 that a cell in the dark acts as a diode. The shading of one cell—converting it into a diode under reverse bias—therefore eliminates the current produced by the entire string. Furthermore, the shaded cell will dissipate all the power produced by the illuminated cells in the string which can be considerable if the string is large.

Fig. 4.7 The hot spot formation

The common technique used to alleviate this effect is to employ *by-pass diodes* which are connected across a block of several cells in a string (Fig 4.6). This limits the power which is dissipated in this block and provides a low-resistance path for the module current.

4.3.4 Capturing the sunlight

An important problem that confronts the designer of an array is whether the modules are to be mounted at fixed positions, or their orientations will follow ('track') the motion of the Sun.

In most arrays, the modules are supported at a fixed inclination facing the equator. This has the virtue of simplicity—no moving parts and low cost. As we have seen in Chapter 2, the optimum angle of inclination depends mainly on the latitude, the proportion of diffuse radiation at the site, and the load profile.

By mounting the array on a two-axis tracker, up to 40% more of the solar energy can be collected over the year as compared with a fixed-tilt installation. But this increases complexity and results in lower reliability and higher maintenance costs. Single-axis tracking is less complex but yields a smaller gain. Where labour is available, the orientation may be manually adjusted to increase the output. It has been estimated that, in sunny climates, a flat plate array moved to face the sun twice a day and given a quarterly tilt adjustment can intercept nearly 95% of the energy collected with a fully automatic two-axis tracking.*

Tracking is particularly important in systems which operate under concentrated sunlight. The structure of these systems ranges from a simple design based on side booster mirrors, to concentration systems which employ sophisticated optical techniques to increase the light input to the cell by several orders of magnitude. These systems must make allowance for an important fact that concentrating the sunlight reduces the angular range of rays that the system can accept for conversion. Tracking becomes necessary once the concentration ratio exceeds about 10 and the system can only convert the direct component of solar radiation.

In this chapter, we shall follow the established practice and lay emphasis on systems with flat panels positioned at a fixed orientation. Concentration systems will be discussed in detail in Chapter 7.

Summary

The structure and properties of the photovoltaic generator and its basic building block, the module, have been examined. The electrical characteristics of the generator depend on the ambient temperature and level of irradiation, and we have shown how the output from a PV generator under operating conditions can be determined. The mismatch losses, and measures

---

* F. C. Treble, lecture notes of the Southampton Short Course on Solar Energy Conversion and Applications.

that can be taken to minimise them, have been discussed. The relative merits of the main options regarding the panel orientation—fixed inclination or tracking—have also been considered.

4.4 ENERGY STORAGE

4.4.1 Introduction

Since the solar energy supply is intrinsically variable in time, stand-alone photovoltaic systems usually make a provision for energy storage.

Although a variety of energy storage methods are under consideration (Table 4.1), the majority of stand-alone PV systems today use battery storage. The notable exception are water pumping systems where the storage of pumped water is the preferred alternative (see section 5.4). The batteries in most common use are lead acid batteries because of their good availability and cost effectiveness. Their specific features will be discussed below. Nickel cadmium batteries are used in some smaller applications where their ruggedness, both mechanical and electrical, is considered essential. However, their high cost per amount of energy stored has prevented their wider use in photovoltaics.

Table 4.1 Some energy storage systems

Energy stored	Technology	Remarks
	Pumped water	1. Common utility use as large-scale energy storage 2. PV pumping (Chapter 5)
Mechanical	Compressed air	Demonstrated technology for large-scale storage
	Flywheel	Under investigation for small systems
Electromagnetic	Electric current in superconducting ring	New development potential using 'high-temperature' superconducting materials
Chemical	Batteries	Discussed in this section
	Hydrogen production	See Chapter 7

As we shall see in Chapter 5, future use of PV power to supplement grid electricity is being seriously addressed in many industrialised countries. The grid would then provide the natural and cheapest storage option. In the longer term, hydrogen production, combined with PV power generation, offers the attractive possibility of comprehensive energy technology with minimum environmental risks, as discussed in Chapter 7.

4.4.2 Battery operation in PV systems

Batteries in PV systems operate under specific conditions which must be allowed for in the system design, as they affect both battery life and the efficiency of the battery operation. The most prominent feature is *cycling* with various cycles of different degree of regularity (Fig. 4.8). During the *daily cycle*, the battery is charged during the day and discharged by the

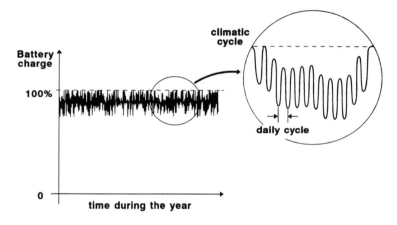

Fig. 4.8 The cyclic operation of battery in PV systems

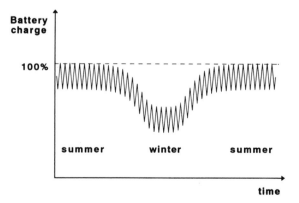

Fig. 4.9 The seasonal cycle

night-time load. The depth of discharge in the daily cycle for systems without back-up is always fairly shallow.

Superimposed on the daily cycle is the *climatic cycle*, due to the variable climatic conditions. This cycle occurs anytime when the daily load exceeds the average energy supply from the PV generator.

In systems where reliability is not the paramount consideration, the battery may act as a *seasonal buffer*. In this instance, the climatic cycle extends over a substantial part of the season (Fig. 4.9).

The details of the cyclic operation depend on the desired reliability of the system, and on the relationship between the storage capacity and size of the PV generator. We return to these points in section 4.6 when we discuss sizing.

4.4.3 Lead-acid batteries

In simplified terms, the lead-acid battery comprises two electrodes of lead and lead dioxide, and the electrolyte of sulphuric acid diluted with water. In practical construction, the electrodes are formed by a lead grid (sometimes alloyed with calcium or antimony) carrying the active material in the form of a porous structure that offers a large surface area for chemical reactions with the electrolyte.

The chemical reactions that take place during battery operation are shown in Fig. 4.10. During the charging process, lead oxide is formed at the anode, pure lead is formed at the cathode, and sulphuric acid is liberated in the electrolyte. During discharge, lead sulphate is formed at both electrodes and sulphur acid is removed from the electrolyte.

discharge to load   **charging by PV generator**

Fig. 4.10 Lead acid battery operation

Fig. 4.11 Battery discharge characteristics under different discharge rates

A typical voltage behaviour during the discharge of a lead-acid battery is shown in Fig. 4.11. It is seen that the battery capacity decreases markedly when discharged at a high rate. For example, a battery with nominal capacity specified at the 10 h discharge rate can markedly increase its capacity under 100 h discharge—the rating value which is used in many photovoltaic applications. This characteristic must be taken into account when designing the energy storage of a PV system since many operate under different conditions than specified by the battery manufacturers.

The voltage behaviour during charging is shown in Fig. 4.12. After a relatively slow increase up to about 2.35 V per element, there is a steep

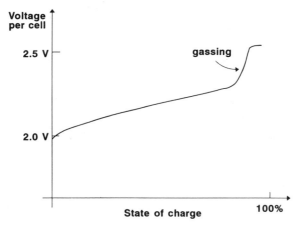

Fig. 4.12 Battery charging characteristic

voltage rise accompanied by *gassing*—the generation of hydrogen and oxygen at the electrodes.

These chemical reactions imply various deleterious processes that accompany the battery operation and shorten battery life or increase maintenance requirements.

Gassing increases the need for maintenance and may represent a safety hazard. In moderate levels, it can be used to advantage by alleviating stratification of the electrolyte (see below).

The repeated growth and dissolution of lead, lead oxide and lead sulphate with different specific volumes is accompanied by mechanical stresses at the electrodes. These stresses, in turn, result in *shedding of active matter* from the electrodes. Another instance where this phenomenon may occur is during intense gassing.

Sulphation is a formation of large lead sulphate crystals at the plates which hinder the reversible chemical reactions. This phenomenon occurs mainly when the battery remains in a low charge state for extended periods of time.

The battery operation tends to favour a non-uniform electrolyte distribution where the electrolyte with the highest density occurs at the bottom of the battery vessel. This *stratification* of the electrolyte promotes corrosion and sulphation of the bottom part of the negative electrode, but can be avoided by a regular weak overcharge in which gassing is used to stir the electrolyte.

It is a good practice to give the batteries an *equalising charge* at the end of a period when the battery will have operated in low charge state (normally, at the end of winter). This boost charge which briefly overcharges the battery ensures that all the elements are fully recharged.

Electrode corrosion occurs particularly by the growth of the positive lead grid, and is accelerated by elevated temperatures.

Table 4.2 Principal types of commercial lead-acid batteries

Type	Application	Operation	Characteristics
SLI	Vehicle starter batteries	Supply of high peak power for short time	Low resistance to cycling Low cost
Traction	Electric vehicles	Deep and frequent cycles	High resistance to cycling High water consumption Frequent maintenance
Stationary storage	Emergency power supplies	Float operation in high state of charge	Moderate resistance to cycling Low water consumption

The design of the battery storage must also ensure that the electrolyte does not freeze during operation at sites where the temperature can drop to low levels. This can be avoided by maintaining a relatively high density of the electrolyte, either by high state of charge or by increasing the electrolyte density above the usual norm.

Although specialised PV batteries are now becoming available on the market, most batteries that are currently installed in PV systems are standard components originally intended for conventional application, or adapted from them to suit the particular mode of operation envisaged for the PV system. The three principal classes of lead-acid batteries are listed in Table 4.2.

Summary

Lead-acid batteries provide the most common means of energy storage in PV systems today. A prominent feature of their operation is cycling. This, together with other operation parameters, affect the battery life and maintenance requirements which must be allowed for in the design of a PV system.

4.5 POWER CONDITIONING AND CONTROL

Various electronic devices are used to accomodate the variable nature of power output from the PV generator, to avoid the malfunction of the system, or to convert the DC power produced by the PV generator into AC output.

4.5.1 The blocking diode

We have seen in Chapter 3 that a solar cell in the dark behaves as a diode. Similarly, the characteristic of a photovoltaic generator at night can be obtained by displacing its usual characteristics under illumination along the current axis, until it passes through the origin. Without special precautions, this type of night-time operation of the photovoltaic generator will provide a discharge path for the battery. The simplest solution is to separate the generator and battery by a *blocking diode* (Fig. 4.13). When the voltage at the battery exceeds the voltage at the generator, the diode becomes reverse-biased and prevents the battery discharge. During daytime operation, however, there will be a voltage drop across the blocking diode which should be taken into account when designing the system.

In systems using modern PV modules where the series resistance is low and the I–V characteristic approaches the ideal curve, the battery discharge

Fig. 4.13 The operation of the blocking diode (after E. Lorenzo, *Electricidad Solar Fotovoltaica*, publication of E.T.S.I. Telecomunicacion, Madrid, 1984)

current via the PV generator at night can be very small. The power dissipated at the blocking diode during daytime operation may then exceed the night-time discharge losses. For this reason, the blocking diode is sometimes omitted from the circuit design.

4.5.2 Self-regulating systems

In small systems, the connection of the battery with the PV generator via a blocking diode provides the simplest solution. This configuration relies on a correct choice of the operating point of the PV generator to match the battery charging requirements (Fig. 4.14). The usual operation of the system takes place between the battery voltage under the minimum allowed state of charge, and the voltage at full charge, allowing for a voltage drop across the blocking diode. If the array operating voltage is set at the upper end of this voltage range, a slight increase in the battery voltage then sharply reduces the charging current from the PV generator and prevents battery overcharging.

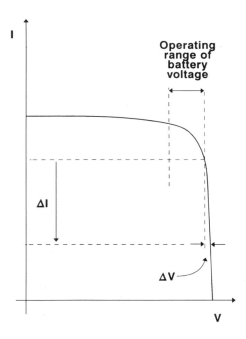

Fig. 4.14 The operation of self-regulating PV system

We should emphasize, however, that the simplicity of design in these systems suffers serious drawbacks in practical operation. The fact that there is no battery discharge protection can have a very detrimental effect on the battery life. The system design is very sensitive to the ambient temperature and, in many circumstances, would yield very inefficient performance. For these reasons, self-regulation is usually recommended only when system costs must be kept at the absolute minimum.

4.5.3 Charge regulator

As discussed in section 4.4, measures must be taken to prevent excessive discharge and overcharge of batteries. Various types of charge regulators

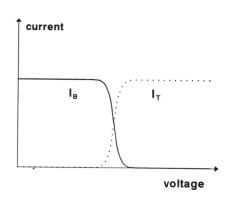

Fig. 4.15 The operation of shunt regulator (after E. Lorenzo, *Electricidad Solar Fotovoltaica*, publication of E.T.S.I. Telecomunicacion, Madrid, 1984)

are available that fulfil this role. In small applications (up to about 100 W), a *shunt regulator* can be used to dissipate the unwanted power from the generator. A common implementation is to use a transistor in parallel with the PV generator (Fig. 4.15) which is set to conduct and divert current from the battery at a certain threshold voltage value.

In larger applications, it is advisable to disconnect the battery from the generator by means of a *series regulator*. This can be an electromechanical switch (for example, a relay) or a solid-state device (bipolar transistor, MOSFET, etc.). The former devices have the advantage that they do not dissipate energy but their reliability can be a problem in locations with high dust or sand occurrence.

The battery may be protected against excessive discharge by a *charge limiter*. This device is introduced between the load and the battery, and acts as a switch which opens when the battery charge reaches a minimum acceptable level.

4.5.4 DC/DC converter

The variability of the power output from the PV generator implies that, without special interface measures, the generator will often operate away from its maximum power point. The associated losses can be avoided by the use of a *maximum-power-point tracker* which ensures that there is always a maximum energy transfer from the generator to the battery.

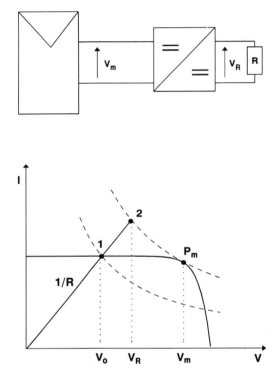

Fig. 4.16 The operation of MPP tracker (after E. Lorenzo, *Electricidad Solar Fotovolta-ica*, publication of E.T.S.I. Telecomunicacion, Madrid, 1984)

The principles of the MPP tracker are demonstrated in Fig. 4.16 for the situation when the PV generator feeds power to a resistive load R. Shown are the I–V characteristics of the generator and the load, together with constant power curves $P = IV =$ constant. It is seen that at the operating point 1 (which corresponds to a direct connection between the generator and the load) the delivered power is significantly below P_{max}, the maximum power of the PV generator. A DC/DC converter will transform the voltage at the load to a value

A. Variable frequency inverter drive
   system without battery

B. Fixed frequency self-commutated inverter
   system with battery

C. PV power plant (line-commutated inverter)

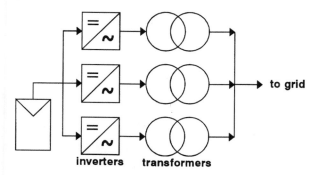

D. Large multiple inverter PV power plant

Fig. 4.17 Various inverter configurations

$$V_R = \sqrt{P_{max}R} \qquad (4.6)$$

thus ensuring a maximum power transfer.

The use of a MPP tracker is usually justified in systems with peak power in excess of 1 kW$_p$.

4.5.5 DC/AC converter (inverter)

This is a standard item of electronic equipment which is used in many different applications. The input power is the DC power from the photovoltaic generator or battery, and the output is AC power used to run AC appliances or fed into the utility grid. The efficiency of the inverters usually depends on the load current being a maximum at the nominal output power. It can be as high as 95% but will be lower (75–80%) if the inverter runs under part load

Table 4.3 Inverter parameters

	Line-commutated inverters	Self-commutating PWM inverters
High unit power availability	Yes	No
High-voltage operation	Yes	No
Cost	Low	High
Reliability	High	Low
Ability to operate stand-alone (during grid black-out)	No	Yes
Harmonic content	High	Low
Need for power factor correction	Yes	No
Development status	Technology standard	Require R&D
Space required	Large	Small

The majority of inverters for PV applications can be classified into three main categories (Fig. 4.17). *Variable frequency inverters* are used for stand-alone drive/shaft power applications, almost exclusively in PV pumping systems. The remaining two main inverter categories are suitable for the grid connection of PV power plants. *Self-commutating fixed-frequency inverters* are able to feed an isolated distribution grid and, if equipped with special paralleling control, also a grid supplied by other parallel power sources. *Line-commutated fixed-frequency inverters* are able to feed the grid only where the grid frequency is defined by another power source connected in parallel. The inverter will not work if such external frequency reference is lacking (e.g. during grid black-out). The advantages and drawbacks of these two inverter types are summarised in Table 4.3.

4.5.6 Alarms, indicators and monitoring equipment

The system electronics should include some indicators which display the state of the system, or at least its main parameters. The main indicators should display the low charge state for the batteries and the overcharge.

In some instances, the user should be warned about the state of the system by an alarm. An example of this situation could be when the battery charge is low, and the charge limiter is about to cut off the supply.

The operation of many photovoltaic systems is monitored and the values of various parameters registered. For example, it is useful to have information about the energy collected and supplied, and about the irradiance, temperature and voltage produced by the system.

Summary

The power conditioning and control elements make it possible to convert the generated DC power to AC, protect the battery against overcharge or excessive discharge, and optimise the energy transfer between the PV generator and the battery or load. Indicators and monitoring equipment were also briefly discussed.

4.6 SIZING PHOTOVOLTAIC SYSTEMS

4.6.1 Introduction

Sizing of a PV system, particularly a stand-alone one, is an important part of its design. Since the capital equipment cost is the major component of the price of solar electricity, oversizing the plant has a very detrimental effect on the price of the generated power. Undersizing a stand-alone system, on the other hand, reduces the supply reliability.

The sizing of a system requires a knowledge of the solar radiation data for the site, the load profile, and the importance of supply continuity. In addition, other constraints on the design (for example, economic) must also be known. The sizing procedure then recommends the size of the photovoltaic generator and battery capacity that will be optimum for the application. It will also allow the nominal characteristics of the electronic components to be specified.

4.6.2 Radiation and load data

Radiation data for a particular site (when available) are usually given in the form of global irradiation on a horizontal surface. We have shown in

Chapter 2 how this information, in the form of monthly averaged daily data, can be translated into the radiation received by a panel at an inclined orientation (Table 2.2 and Figs. 2.9, 2.10). Some of these results are reproduced in Table 4.4.

Table 4.4 Daily irradiation in Barcelona (data in kWh/m², panel inclination 60°)

January	4.60	July	4.80
February	4.59	August	4.85
March	4.62	September	4.58
April	4.55	October	4.53
May	4.30	November	4.12
June	4.45	December	3.49

Annual mean 4.46

The *load data* give detailed information about the appliances or equipment to be powered: their number, nominal power, nominal operating voltage and the number of hours of operation in a typical day.

To imagine what a typical load requirement might be, we give here some examples for home appliances. The appliances in question may be lights, washing machine, refrigerator, water pump, TV set, or other electric appliances. Every appliance can be described by its nominal power, voltage, and hours of operation in a typical house. Some of these data are summarised in Table 4.5.

Table 4.5 Data for typical loads

Appliance	Nominal power (W)	Nominal voltage (V)
Light appliance 1	15	24
Light appliance 2	20	24
Washing machine	100/800	220 rms
Refrigerator	70/150	220 rms
Small electric appliances	250	220 rms
Water pumping	100	24
TV set	60/100	24
Others (radio emitter/receiver)	120	24

The size of the PV installation depends on the number and hours of operation of each of these appliances. One can distinguish various categories according to the energy demand per day as shown in Table 4.6.

Type A is a simple system providing energy to power five light appliances (at an average of 1.65 hours per day), a TV set in operation for four hours per day, and the use of small electrical appliances.

Type B is a small but sufficient installation to power 12 light appliances, a DC washing machine (2 kg) with cold water, small electrical appliances, TV and water pumping.

Table 4.6 Typical loads for housing

System type	Energy supplied (Wh/day)	Charge at 24 V (Ah/month)
A	500	625
B	1000	1250
C	1500	1875

Type C is a size more adequate for housing and small agricultural needs. More energy is therefore allocated for water pumping and more light appliances. A washing machine (220 V AC) can also be powered.

In addition to the average power demand of the load, the *load profile*—both annual and daily—also has an important effect on the design of the system. Clearly, a system powering a load whose peak matches periods of high solar radiation will have smaller storage requirements. In some systems, the load can be adjusted to respond to the energy input to the system. An example can be a stand-alone domestic system where a washing machine is allowed to be used only when there is enough sun.

4.6.3 The system energy balance

A relatively simple sizing procedure can be adopted for some, usually attended, systems where supply reliability is not of paramount importance. The system design can then be based on the yearly energy balance between the radiation and the load.

We shall show how this sizing method is implemented for an array with a fixed panel orientation (Fig. 4.18). The panel inclination can be chosen according to the general rules which were discussed in Chapter 2, or it can be treated as a parameter subject to optimisation in the final design.

4.6.3.1 Input to the sizing procedure

[Blocks 1–3] *Determination of the energy input*. The radiation data for the site, together with the panel orientation, are used to determine the incident solar radiation on the panel for a typical day in every month of the year (see section 2.2.3).

[4, 5] *Determination of the load demand*. The load specification, or the typical load for a similar system, is used to determine the load power demand on a typical day.

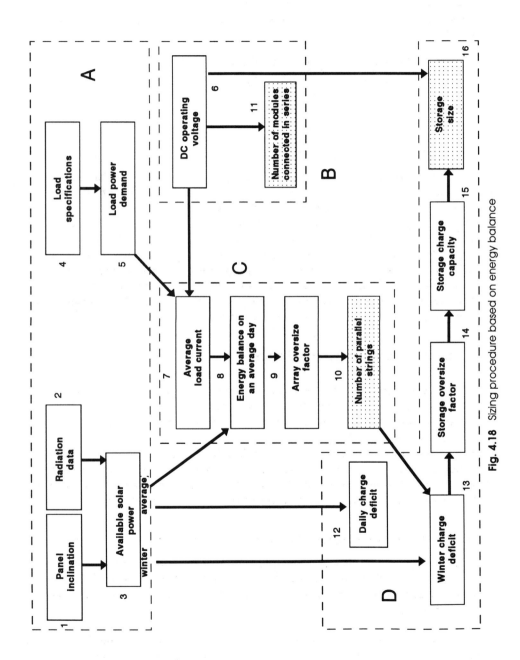

Fig. 4.18 Sizing procedure based on energy balance

4.6.3.2 Number of series-connected modules

[6] The DC operating bus bar voltage V_{DC} of the system is specified.

[11] As we have seen in section 4.3, the number of modules N_s which are to be connected in a series string is directly determined by the DC operating voltage. We can immediately write down

$$N_s = \frac{V_{DC}}{V_m} \qquad (4.7)$$

where V_m is the operating voltage of one module which was discussed in section 4.3.

4.6.3.3 The number N_p of parallel strings

This number is directly related to the current requirement of the load.

[7] The equivalent load current is calculated from the equation

$$I_L(A) = \frac{E_L}{24\,V_{DC}} \qquad (4.8)$$

where E_L (Wh/day) is the typical power requirement for the load.

[8] We now define the nominal current I_p which is required from the photovoltaic generator when working at its maximum power point and irradiated by the standard AM1.5 radiation at 1 kW/m². The energy balance for a typical day can then be written as:

$$E_L(\text{Wh/day}) = \text{PSH}\, I_p V_{DC} \qquad (4.9)$$

where we have expressed the radiation incident on the panel in *peak solar hours*, equal to the number of hours of the standard irradiance (1 kW/m²) which would produce the same irradiation. Clearly, PSH is numerically equal to the irradiation in kWh/m² day. Using equation (4.8), equation (4.9) becomes

$$I_p = \frac{24\,I_L}{\text{PSH}}. \qquad (4.10)$$

Equation (4.10) has a simple interpretation. The average load current, multiplied by the number of hours in the day, must be equal to the nominal (peak) current from the PV generator, multiplied by the number of peak solar hours.

[9, 10] The number of modules to be connected in parallel is then calculated using the following equation:

$$N_p = (SF)\, \frac{I_p}{I_m} \tag{4.11}$$

where I_m is the nominal current supplied by an individual photovoltaic module when illuminated under standard conditions and SF is a *sizing factor* which is introduced to oversize the amount of current available from the array. We shall see below how the value of SF affects the overall design of the system.

4.6.3.4 Sizing of the storage subsystem

[12] The daily and seasonal charge deficits are now calculated. It must be ensured that the night periods or shadows are covered satisfactorily. At the same time, excess energy generated and not used during the sunshine periods must be stored. This analysis determines the daily charge/discharge percentage of the battery that usually cannot exceed a given value for safe operation.

[13] The energy balance for the year is set in such a way that the summer energy excess can be stored effectively to cover the cumulative energy deficit during the winter period. Thus, one calculates the winter energy deficit ΔE due to the seasonal weather cycle. This can be expressed as a charge deficit Q_{yd}, usually given in ampere-hours,

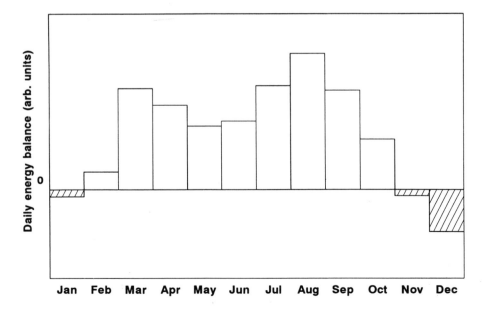

Fig. 4.19 The daily energy balance of a stand-alone PV system

$$Q_{yd} = \frac{\Delta E}{V_{DC}}. \tag{4.12}$$

This result is shown in Fig. 4.19 where the value of ΔE is represented by the shaded area. Note that this energy deficit depends on the choice of the array sizing factor SF.

[14] A further charge deficit Q_{los} is now added to allow for a number of days of operation without energy input (because of lack of sunshine, system maintenance, or servicing). This number is here determined from experience, and depends on the particular site of the system. The total charge deficit to be covered by the storage subsystem is then determined, together with the application of pertinent safety factors.

[15] One can now calculate the nominal capacity of the battery:

$$Q_B = (Q_{yd} + Q_{los}) \, (1/\Phi) \, (Ah) \tag{4.13}$$

where Φ is a factor that accounts for the fact that the batteries cannot be completely discharged. As we have seen in section 4.4, the value of this factor has an important effect on the battery life.

[16] Given the DC operating voltage and manufacturer's specifications, one can determine the required size of the storage subsystem. This completes the preliminary system design.

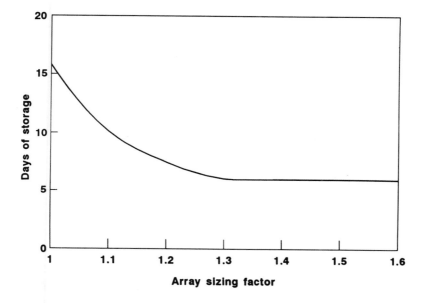

Fig. 4.20 The storage–array relationship based on energy balance

One may, at this point, examine the free variables that can be optimised. For example, the effect of the panel sizing factor SF on the battery storage capacity should be examined (Fig. 4.20). Using this graph, one can choose the optimum array size and storage to minimise the cost of the system. Similarly, one can also optimise the panel orientation.

Once the storage capacity and panel size have been selected, it is recommended to simulate numerically the operation of the system, and to determine the reliability of the power supply. An example of such procedure is shown in Fig. 4.21 (adapted from Macomber *et al.*, 1981).

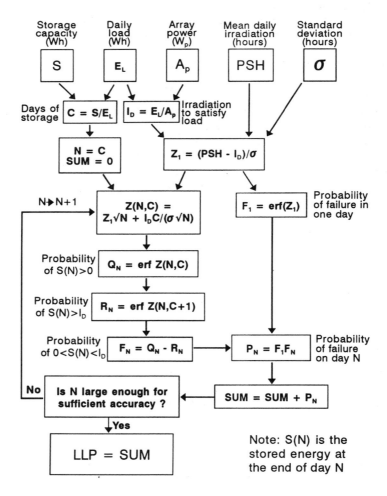

Fig. 4.21 Algorithm to calculate the reliability of PV systems (after H. L. Macomber *et al.*, *Photovoltaic Stand-Alone Systems: Preliminary Engineering Design, Handbook* prepared for NASA Le RC DEN 3–195, 1981)

The inputs to this procedure are the system parameters (the selected storage size, average daily load, nominal array power) and the radiation data for the site which can be found in tables (the mean daily irradiation—expressed here in peak solar hours—and the standard deviation from the mean). The procedure calculates the supply reliability in terms of the *loss-of-load probability*, or the relative fraction of time that the supply is likely to be absent. We shall discuss this important concept in section 4.6.4.

4.6.3.5 Worked example

In this example which illustrates the sizing method just described, the load current I_L has been set equal to 1 A. The calculation is then particularly transparent, and the design is quite general since the the current I_p required from the PV panels scales linearly with I_L.

The calculation is carried out using the irradiation data for Barcelona which are given in Table 4.4. Let us suppose that we choose the panel inclination $\beta = 60°$. This is close to the value which maximises the energy input during the 'worst month' (here, December).

The average value for PSH from the data in Table 4.4 is 4.46 h and using equation (4.10) we obtain:

$$I_p = 5.38 \text{ A} \tag{4.14}$$

The number of parallel modules is calculated using equation (4.11),

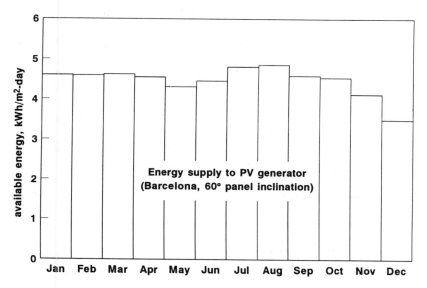

Fig. 4.22 Energy supply to photovoltaic generator

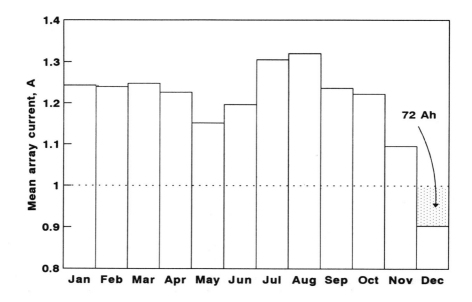

Fig. 4.23 The mean current produced by the photovoltaic generator over the year

$$N_p = \frac{5.38}{2.1} = 2.56 \qquad (4.15)$$

The nominal current of one module was assumed here equal to 2.1 A.

N_p is now set to 3. This gives an effective safety factor of SF = 1.17, and an effective oversizing of 17% in panel current.

Fig. 4.22 shows the incident energy on the panels as a function of the month of the year. The mean value is 4.46 kWh/m² day.

Fig. 4.23 shows the resulting current supplied by the three parallel panels. It is clear that the year mean value is not 1 A but 1.17 A as a result of the effective oversizing.

In the example, the winter deficit Q_{yd} is 72 Ah. Assuming a maximum period of 5 days without sunshine, $Q_{los} = 5 \times 24 \times 1$ Ah = 120 Ah. A further charge capacity (a fraction of a day's load) could be added to compensate for an unfavourable daily load profile. Finally, assuming a conservative value of 70% maximum discharge, the capacity needed for the battery is set at 647 Ah. When choosing the battery one should remember that this capacity is under 100 h discharge rate.

4.6.4 Sizing and reliability

In many applications of photovoltaics today, reliability of supply is an important factor, and must be reflected in the sizing calculation. It is then highly desirable that the procedure gives, as a final result, the cost effective combination of the array size and storage capacity, subject to a minimum acceptable level of supply reliability.

The supply reliability is measured by the *loss-of-load probability* (LLP), defined as

$$LLP = \frac{\text{Average duration of supply interruption}}{\text{Total duration of supply}}.$$

Some examples of the LLP values for different applications are given in Table 4.7.

Table 4.7 Recommended loss-of-load probability values

Application	LLP
Domestic:	
Illumination	10^{-2}
Domestic appliances	10^{-1}
Telecommunications	10^{-4}

As we have seen in section 4.3, the design of a photovoltaic system can be verified to satisfy the necessary reliability requirements by calculating the LLP figure for the design. Methods are becoming available which allow one

Fig. 4.24 Sizing based on loss-of-load probability

to input this information to the sizing procedure from the start, and generate a set of *isoreliability lines* (Fig. 4.24. Egido and Lorenzo, 1992). Each line is a result of sophisticated computer algorithm, and represents the combination of storage capacity and array area which satisfies the reliability level described by the LLP value. These curves can then be further analysed to find the optimum combination with respect to cost, and represent the ideal information needed for a system design. Since this type of analysis is, at present, available only for a limited number of sites, more intuitive methods are often used to install high-reliability PV systems in practice.

Summary

The sizing calculation determines the size of the array and battery for a specific application, and must take into account the following factors:

(a) Power supply from the sun and the requirement from the load.
(b) Maximum acceptable level of supply interruption, or loss-of load probability (LLP), to allow for the random variation of supply and load.
(c) Cost and other constraints relating to the system.

A sizing procedure based on the energy balance in the system was considered in detail, and a more rigorous method of based on LLP outlined.

SUMMARY OF THE CHAPTER

The photovoltaic system usually consists of a number of subsystems. In addition to the photovoltaic generator, provision is usually made for energy storage and power conditioning, including control elements. Some systems also have a back-up generator.

The construction of a PV generator from modules has been discussed, focusing attention on the electrical characteristics of the generator under operating conditions and problems which arise from module interconnection.

The most common method of energy storage in PV systems is by lead-acid batteries. The characteristics of their operation have been analysed, alongside various chemical and physical processes which limit battery life or increase the maintenance requirements.

Power conditioning and control elements are often included to feed AC loads or the grid, interface different parts of the system, or monitor the system performance.

Sizing is an important part of the system design, particularly for stand-alone systems. The sizing procedure uses the radiation and load data to recommend the sizes of the array and battery storage, subject to the required reliability of power supply. It also allows the system cost to be minimised.

BIBLIOGRAPHY AND REFERENCES

EGIDO, M. A. and LORENZO, E., The sizing of stand-alone PV systems: a review and a proposed new method, *Solar Energy Materials and Solar Cells* **26**, 1992: 51–69.
KALHAMMER, F. R., Energy-storage systems, *Scientific American* **241** (6), 1979: 42–51.
LORENZO, E. *et al.*, *Electricidad Solar Fotovoltaica*, E.T.S.I. Telecomunicacion, Madrid, 1984.
MACOMBER, H. L. *et al.*, *Engineering Design Handbook for Stand-Alone Photo-voltaic Systems*, NASA contract DEN 3–195, 1981.

SELF-ASSESSMENT QUESTIONS

Part A. True of false?

1 The hot-spot formation arises from non-uniformity of module encapsulant.

2 The blocking diode is connected in parallel with a string of modules.

3 Deep discharge improves the life of the lead-acid battery.

4 When modules are connected into a string, they should be selected according to similar short-circuit current.

5 The maximum power point of the module specified by the manufacturer determines the operating point of the battery.

6 The battery voltage increases when the load current is reduced.

7 The photocurrent depends exponentially on the temperature.

8 A DC/DC converter improves the interconnection between the AC/DC converter and the load.

9 Periodic overcharging of the battery improves the homogeneity of the electrolyte.

10 Once the LLP value is fixed, the relationship between the array size and storage size is described by a curve which resembles a hyperbola.

PART B

1   What subsystems of a PV system can you name?

2   What are the main categories of PV systems?

3   How do the current and voltage of a module depend on the operating conditions?

4   How many cells usually comprise a module? How are the cells connected and why?

5   What is the role of a by-pass diode?

6   What are mismatch losses and how can they be reduced?

7   What is the temporal pattern of battery operation in a PV system?

8   Why is it a good maintenance practice to overcharge the battery at the end of winter?

9   Is it true to say that the operational capacity of a battery in a PV system can be significantly higher than the manufacturers specifications?

10   What are the disadvantages of operating battery storage as a seasonal buffer?

11   What is the role of the blocking diode? Is it always necessary?

12   How can a PV system operate without electronic control and what disadvantages does this system design have?

13   What types of charge regulators are used in PV systems?

14   Which inverter would you use in a stand-alone domestic system?

15   Which control element would you use to ensure maximum power extraction from PV generator?

16   Why is accurate sizing of PV installations important?

17   What data are needed to size a PV system?

18   What is a measure of reliability of electricity supply of a PV system?

19   Why is the panel inclination in a stand-alone system usually chosen steeper than in a grid-connected one?

20   Which parts of the block diagram in Fig. 4.18 are determined by the required reliability of the system? How does the design of the PV generator depend on the chosen DC bus-bar voltage? How does it depend on the required current?

Answers

Part A

1, False; 2, False; 3, False; 4, True; 5, False; 6, True; 7, False; 8, False; 9, True; 10, True.

PART B

1 PV generator, storage, power conditioning, control and possibly a backup generator.
2 Stand-alone and grid-connected systems
3 The principal factors are voltage decrease with increasing temperature and current proportional to the irradiance.
4 Between 33 and 36 cells connected in series to charge a 12 V battery.
5 The by-pass diode protects the modules from hot-spot formation and reduced output losses on partial shading or damage to the array.
6 Mismatch losses occur as a result of variable module or cell quality and because of partial shading of the array. They can be reduced by sorting the arrays prior to construction and by installing by-pass diodes.
7 Cycling with daily and climatic cycles, possibly with a seasonal cycle.
8 To provide an equalising charge for the battery elements.
9 Yes. The battery capacity under slow discharge can be considerably higher than under the discharge rate specified by the manufacturer.
10 Long period of operation in low charge state may lead to electrolyte stratification and sulphation.
11 To prevent the array discharging the battery at night. It should be omitted when the power dissipation at the diode during the day exceeds the energy loss by the discharge at night.
12 By setting the array operating voltage to fully charge the battery but in such a way that a further slight voltage increase produces a large drop in the charging current. The operation of such system is unlikely to be completely satisfactory as is does not protect the battery against excessive discharge, and the power transfer from the generator will not be optimum at other temperature than that considered in the design.
13 Shunt (parallel) and series regulators.
14 Self-commuted inverter.
15 Maximum-power-point tracker.
16 Undersizing reduces the supply reliability and oversizing increases the cost.
17 Load and radiation data, and the required supply reliability.
18 Loss-of-load probability, LLP.
19 To maximise the energy collection in winter, thus reducing the storage requirements.
20 (a) Blocks 9 and 14. (b) The DC bus-bar voltage determines the number of cells connected in series in each string. (c) It determines the number of parallel strings.

5

Applications

AIMS

The aim of this unit is to introduce the most common applications for solar
electricity.

OBJECTIVES

After completing this unit the student should be able to:

1. Name a wide range of applications for photovoltaics.
2. Explain the basic features of each application.
3. Understand why photovoltaics is a suitable option in each case.
4. Assess the economic viability of a PV installation.

NOTATION AND UNITS

Symbol or acronym		SI unit	Other unit
A	Array area	m^2	
ALCC	Annualised life-cycle cost		
Ca	Annually recurring payment		
Cr	Single future payment		
d	Discount rate		
E	Daily subsystem efficiency		
E_m	Module efficiency		
E_s	System efficiency		

Edited from manuscripts of O. Paish, B. McNellis and A. Derrick (Overview of PV applications,
Economics, Water pumping), M. Hill and S. McCarthy (Domestic PV supply), B. Hill (Remote
monitoring and control) and K. Bogus (Electric power generation in space). Additional
information by A. Sorokin.

115

(*Contd.*)

Symbol or acronym		SI unit	Other unit
F	Array mismatch factor		
i	Excess inflation rate		
I	Yearly mean daily irradiation	J/m^2	Wh/m^2
N	Number of years		
Pa	Discount factor		
Pr	Discount factor		
PW	Present worth		

5.1 INTRODUCTION

There is an extensive range of applications where the PV system is already viewed as the best option for electricity supply (Fig. 5.1). These systems are usually stand-alone, and exploit the following advantages of PV electricity:

- There are no fuel costs or fuel supply problems.
- The equipment can usually operate unattended.
- It is very reliable and requires little maintenance.

The reason for PV success can be purely commercial, based on conventional economic arguments which will be discussed in section 5.2. Here belong the so-called professional applications (reviewed in section 5.5), and a wide range of consumer products such as electronic watches, calculators, power for leisure equipment and tourism, etc.

Table 5.1 PV applications in demonstration and THERMIE programmes of the European Community during 1979–90 (source: Directorate General for Energy, Commission of the European Communities)

Application	No. of projects	%
Houses and villages	42	43.8
Farms and agriculture	10	10.4
Nature reserves and recreation	8	8.3
Lighthouses and buoys	5	5.2
Water pumping and treatment	10	10.4
Warning and monitoring systems	12	12.4
Telecoms and public lighting	9	9.4
Total	96	100.0

In satellite power systems (discussed in section 5.6) where price of the PV system is not the deciding factor, solar cells have been employed for some time by virtue of their superior technical merits, principally their reliability and low power/weight ratio.

An ever growing number of PV systems, however, are installed for their social benefit to rural communities in both the developing and industrial countries of the world. Various aspects of rural electrification will be discussed in section 5.3, with PV water pumping highlighted in section 5.4.

There has been much discussion in recent years about the possibility that photovoltaics will eventually be cheap enough to be economic for grid-connected applications, competing with conventional power stations. We shall discuss two possible scenarios, central and distributed generation, in section 5.7.

An idea about the distribution of PV applications across the various sectors can be obtained from the statistics of the Demonstration and THERMIE programmes of the European Community (Table 5.1).

5.2 ECONOMICS OF PV INSTALLATIONS

5.2.1 Introduction

The aim of this lesson is to learn how to calculate the economic value of a photovoltaic system so that it can be compared with other methods of power generation. The economic viability of a PV system must be assessed relative to the alternatives, for example, diesel, extending the grid, etc. At present, PV is most competitive where small amounts of energy are required far from the grid.

The economics of PV systems is different to that of other small power systems:

- The initial expenditure on the equipment (the capital cost) is high.
- There are no fuel costs.
- Maintenance costs are low.
- Reliability is high so replacement costs are low.
- The output of the system depends on its location.

The economic benefits of the system can either be money that has been saved, or revenue that has been collected, by operating the system. As well as economic benefits, there are also social and environmental benefits which should be taken into account when choosing between systems.

There are two ways of looking at the value of an electricity generating system. The economic approach takes the standpoint of the government, and so considers its value to the economy as a whole. It therefore looks at costs which exclude taxes and subsidies. In contrast, a financial assessment is an evaluation from the buyer's point of view. Therefore taxes, subsidies, interest payments on a loan, etc., must all be taken into account. Only economic assessment will be covered here, since that is the most general case.

RURAL ELECTRIFICATION

- lighting and power supplies for remote buildings (mosques, farms, schools, mountain refuge huts)
- power supplies for remote villages
- battery charging stations
- portable power for nomads

WATER PUMPING AND TREATMENT SYSTEMS

PV lighting for a Community Building

- pumping for drinking water
- pumping for irrigation
- de-watering and drainage
- ice production
- saltwater desalination systems
- water purification
- water circulation in fish farms

Floating Solar Powered Pump

HEALTH CARE SYSTEMS

- lighting in rural clinics
- UHF transceivers between health centres
- vaccine refrigeration
- ice pack freezing for vaccine carriers
- sterilisers
- blood storage refrigerators

COMMUNICATIONS

- radio repeaters
- remote TV & radio receivers
- remote weather measuring
- mobile radios
- rural telephone kiosks
- data acquisition and transmission (river levels, seismographs)
- emergency telephones

PV Refrigerator for Vaccine Storage

AGRICULTURE

- livestock watering
- irrigation pumping
- electrical livestock fencing
- stock tank ice prevention

PV Power for a Radio-Telephone System

GRID-CONNECTED APPLICATIONS

- **Distributed applications in buildings**
- **PV power stations**

TRANSPORT AIDS

- road sign lighting
- railway crossings and signals
- hazard and warning lights
- navigation buoys
- fog horns
- runway lights
- terrain avoidance lights
- road markers

SECURITY SYSTEMS

- security lighting
- remote alarm system

PV Powered Navigation Buoy

CORROSION PROTECTION SYSTEMS

- cathodic protection for bridges
- pipeline protection
- well-head protection
- lock gate protection
- steel structure protection

MISCELLANEOUS

- ventilation systems
- camper and recreational vehicle power
- calculators
- automated feeding systems on fish farms
- solar water heater circulation pumps
- path lights
- yacht/boat power
- vehicle battery trickle chargers
- earthquake monitoring systems
- battery charging
- fountains
- emergency power for disaster relief
- aeration systems in stagnant lakes

PV Street Lamp

INCOME GENERATION

- battery charging stations
- TV and video pay stations
- village industry power
- refrigeration services

PV Powered Ventilator

ELECTRIC POWER FOR SATELLITES

- Telecommunications
- Earth observation
- Scientific missions
- Large space station

Fig. 5.1 Applications of photovoltaics

In an economic evaluation, the following parameters are usually considered:

- *The life-cycle costs.* The sum of all the costs of the system over its lifetime, expressed in today's money.
- *Payback period.* The time it takes for the total costs to be 'paid for' by the monetary profits and other benefits of the system.
- *Rate of return.* The magnitude of the profits and benefits expressed as a percentage annual return on the initial investment.

Payback period and rate of return have two disadvantages. Firstly, it is not always easy to express the benefits gained in monetary terms, and secondly, they do not take account of how long the system will last, nor any future costs that will be incurred as time goes by.

Life-cycle costing is the most complete analysis and is the usual method for determining whether an application is economic. This section will only concentrate on life-cycle costing.

5.2.2 Life-cycle costing

In this method, not just the initial costs, but all future costs for the entire operational life of the PV system are considered. The period for the analysis must be the lifetime of the longest-lived system being compared.

For instance, a PV array costs more to buy than a diesel generator, but the modules should last over 20 years. The diesel generator might last 10 years, using a certain amount of fuel each year. So in this case the analysis period is 20 years. In addition to the capital cost, the cost of a replacement diesel after 10 years, plus 20 years' worth of fuel must also be included for the diesel option. In addition, the costs of maintenance and repair for the two systems over the whole 20 year cycle must be included. Depending on the exact figures, either the PV or the diesel system will work out cheaper overall.

To make a meaningful comparison, all future costs and benefits have to be discounted to their equivalent value in today's economy, called their 'present worth' or PW. To do this, each future cost is multiplied by a discount factor calculated from the discount rate. A discount rate of 10% per year would mean that in real terms it makes no difference to a farmer whether he has $100 now or $110 dollars in one year's time. Therefore a cost of $110 dollars one year from now has a 'present worth' of $100.

5.2.2.1 Parameters

The calculation of life-cycle costs requires values to be known for the following:

- *Period of analysis.* The lifetime of the longest-lived system under comparison.
- *Excess inflation* (i). The rate of price increase of a component above (or below) general inflation (this is usually assumed to be zero).
- *Discount rate* (d). The rate (relative to general inflation) at which money would increase in value if invested (typically 8–12%).
- *Capital cost.* The total initial cost of buying and installing the system.
- *Operation and maintenance.* The amount spent each year in keeping the system operational.
- *Fuel costs.* The annual fuel bill.
- *Replacement Costs.* The cost of replacing each component at the end of its lifetime.

5.2.2.2 Calculation of present worth

There are two types of calculation that are used in life-cycle costing when expressing a future cost or benefit as its present worth. The first is used to calculate the present worth of a single payment, say the replacement of a battery after five years. The second is used to calculate the total net present worth of a recurring cost, such as annual fuel or maintenance costs. This is the sum of many discounted single payments over the analysis period of N years. To avoid the need for complex equations, the relevant PW can be found by multiplying the actual cost by a factor that can be found in Tables 5.2 and 5.3. The formulae used to calculate the tables are:

$$\text{Pr} = \left(\frac{1+i}{1+d}\right)^N$$

$$\text{Pa} = \left(\frac{1+i}{1+d}\right)\left(\left|\frac{1+i}{1+d}\right|^N - 1\right) \Big/ \left(\frac{1+i}{1+d} - 1\right)$$

Single payment. For a single future cost Cr, payable in N years time, the present worth is given by

$$\text{PW} = \text{Cr} \times \text{Pr}.$$

Example. It is estimated that a new pump will be required for a certain solar pumping system in 10 years time. We will assume that a new pumpset presently costs $1000, that pump prices do not change relative to general inflation, and that the discount rate is 10%. Using Table 5.2 with $i = 0$, $d = 0.1$ and $N = 10$ (the number of years hence that the payment is to be made), this gives a discount factor Pr of 0.39. Therefore the present worth of this future cost is

$$\text{PW} = \$1000 \times 0.39 = \$390.$$

Table 5.2 Selected values of present worth factor Pr for a cost in *N* years time

Discount Rate (*d*)	Inflation Rate (*i*)	Factor Pr for given number of years				
		5	10	15	20	30
0.00	0.00	1.00	1.00	1.00	1.00	1.00
	0.05	1.28	1.63	2.08	2.65	4.32
	0.10	1.61	2.59	4.18	6.73	17.45
	0.15	2.01	4.05	8.14	16.37	66.21
	0.20	2.49	6.19	15.41	38.34	237.38
0.05	0.00	0.78	0.61	0.48	0.38	0.23
	0.05	1.00	1.00	1.00	1.00	1.00
	0.10	1.26	1.59	2.01	2.54	4.04
	0.15	1.58	2.48	3.91	6.17	15.32
	0.20	1.95	3.80	7.41	14.45	54.92
0.10	0.00	0.62	0.39	0.24	0.15	0.06
	0.05	0.79	0.63	0.50	0.39	0.25
	0.10	1.00	1.00	1.00	1.00	1.00
	0.15	1.25	1.56	1.95	2.43	3.79
	0.20	1.55	2.39	3.69	5.70	13.60
0.15	0.00	0.50	0.25	0.12	0.06	0.02
	0.05	0.63	0.40	0.26	0.16	0.07
	0.10	0.80	0.64	0.51	0.41	0.26
	0.15	1.00	1.00	1.00	1.00	1.00
	0.20	1.24	1.53	1.89	2.34	3.59
0.20	0.00	0.40	0.16	0.06	0.03	0.00
	0.05	0.51	0.26	0.13	0.07	0.02
	0.10	0.65	0.42	0.27	0.18	0.07
	0.15	0.81	0.65	0.53	0.43	0.28
	0.20	1.00	1.00	1.00	1.00	1.00

Annual payment. For a payment Ca occurring annually for a period of *N* years the present worth is:

$$PW = Ca \times Pa.$$

Example. The fuel costs for a particular diesel generator are $50 per year, and it might be assumed that diesel fuel prices will rise at 5% above inflation. Assuming a discount rate of 10% and a length of analysis of 20 years, Table 5.3 (with $i = 0.05$, $d = 0.1$, and $N = 20$) gives a cumulative discount factor Pa of 12.72. The present worth of the diesel fuel costs is therefore:

$$PW = \$50 \times 12.72 = \$636.$$

Table 5.3 Selected values of present worth factors Pa for an annually recurring cost

Discount rate (d)	Inflation rate (i)	Factor Pa for given number of years				
		5	10	15	20	30
0.00	0.00	5.00	10.00	15.00	20.00	30.00
	0.05	5.80	13.21	22.66	34.72	69.76
	0.10	6.72	17.53	34.95	63.00	180.94
	0.15	7.75	23.35	54.72	117.81	499.96
	0.20	8.93	31.15	86.44	224.03	1418.26
0.05	0.00	4.33	7.72	10.38	12.46	15.37
	0.05	5.00	10.00	15.00	20.00	30.00
	0.10	5.76	13.03	22.21	33.78	66.82
	0.15	6.62	17.06	33.51	59.44	164.68
	0.20	7.60	22.41	51.29	107.59	431.39
0.10	0.00	3.79	6.14	7.61	8.51	9.43
	0.05	4.36	7.81	10.55	12.72	15.80
	0.10	5.00	10.00	15.00	20.00	30.00
	0.15	5.72	12.87	21.80	32.95	64.27
	0.20	6.54	16.65	32.26	56.38	151.24
0.15	0.00	3.35	5.02	5.85	6.26	6.57
	0.05	3.84	6.27	7.82	8.80	9.81
	0.10	4.38	7.90	10.71	12.96	16.20
	0.15	5.00	10.00	15.00	20.00	30.00
	0.20	5.69	12.73	21.44	32.22	62.04
0.20	0.00	2.99	4.19	4.68	4.87	4.98
	0.05	3.41	5.16	6.06	6.52	6.87
	0.10	3.88	6.39	8.02	9.07	10.19
	0.15	4.41	7.97	10.85	13.18	16.58
	0.20	5.00	10.00	15.00	20.00	30.00

5.2.2.3 Life-cycle cost (LCC)

For each payment to be made during the lifetime of the system, the present worth can therefore be determined using the discount factors Pr and Pa. The sum of all PW is the total life-cycle cost of the system.

The process could end at this point, but there are two further ways of expressing the life-cycle costs that are more meaningful and easily understood.

5.2.2.4 Annualised life-cycle cost (ALCC)

This is the total LCC expressed in terms of a cost per year. However, the LCC cannot simply be divided by the number of years in the analysis, as this takes no account of the change in value of money due to inflation and interest rates. The LCC must instead be divided by the factor Pa from Table 5.3, found using the chosen discount rate, inflation rate of zero, and a number of years equal to the analysis period. This is really the reverse process of discounting, and the result is expressed in $/year for each system.

Table 5.4 Life-cycle costing calculation sheet

System description: 250 *W Stand-alone domestic PV supply*

Parameters

Period of Analysis = 20 *years*   Excess Inflation $i = 0$ Discount Rate $d = 10\%$

Capital Cost

Hardware:	$ ....1900............	
Installation:	$ .... 400............	
Total:	$ 2300	

Operation and Maintenance:

Annual Cost	$ 50	per year
Discount Factor (Pa)	.... 8.51...........	
Present Worth:	$ 425	

Fuel

Annual Fuel Costs:	$ ....Nil..............	per year
Discount Factor (Pa):	....	
Present Worth:	$ Nil	

Replacements

Item	Year	Cost	Pr	PW
Battery	5	$500	0.62	$310
Battery	10	$500	0.39	$195
Battery	15	$500	0.24	$120
			Total	$ 625

Total Life-Cycle Cost	$ 3350

Annualisation Factor (Pa)	...... 8.51..........
Annualised Life-Cycle Costs	$ 394 per year

Table 5.4 is a calculation sheet for performing a life-cycle costing, filled in for a sample PV system. Table 5.5 is an economic comparison between a PV and a diesel system using life-cycle costing.

Table 5.5 Life-cycle costing comparison: PV vs diesel generating systems

Period of analysis	20.00	years
Discount Rate	10.00	%
Annualisation Factor	8.51	

PV Calculation

Load	2.00	kWh/day	
Battery Efficiency	70.00	%	
Demand at array	2.86	kWh/day	
Days of Battery Storage	5.00	days	
Design Insolation	5.50	kWh/m² day	
Array Mismatch Factor	0.90		
Module Price	4.50	$/$W_p$	
Battery Price	100.00	$/kWh	
Array Size	577	W_p	
Battery Size	10.00	kWh	
			Lifetime
Array Cost	2597.40	$	20 yrs
Battery Cost	1000.00	$	5 yrs
Support/Wiring	866	$	20 yrs
Power Control	50.00	$	10 yrs
Capital Cost	4513.20	$	
Installation	902.64	$	
Total Installed Cost	5415.84	$	
O&M Costs	108.32	$/yr	
Life-Cycle O&M	922.16	$	

Recurring Costs

Array	0.00	$
Battery	1245.86	$
Support/Wiring	0.00	$
Power Control	19.28	$
Total Replacements	1265.13	$

Life-Cycle Cost	7603.14	$
Annualised LCC	893.06	$
Unit Electricity Cost	1.22	$/kW

Diesel Calculation

Using 5 KVA Low Speed Diesel Engine

Load	2.00	kWh/day
		Lifetime
Generator	3000.00 $	20 yrs
Installation	600.00 $	
Installed Capital Cost	3600.00 $	
O&M Costs	360.00 $	
Life Cycle O&M	3064.88 $	
Diesel Price	0.26 $/litre	
Engine Efficiency	25.00 %	
Diesel Consumption	11.00 kWh/litre	
Diesel Usage	0.73 litres/day	
Diesel Cost	69.02 $/year	
Life-Cycle Fuel Cost	587.59 $	

Replacement Costs

Generator	0.00	$

Total LifeCycle Cost	7252.47	$
Annualised LCC	851.87	$
Unit Electricity Cost	1.17	$

5.2.2.5   Unit electricity cost

Probably the most valuable figure for comparing two electricity-generating systems is the net cost of generating each kilowatt-hour during the lifetime of each system. This can be determined from the ALCC as follows:

$$\text{Electricity cost (\$/kWh)} = \frac{\text{ALCC (\$/year)}}{\text{Electricity supplied (kWh/year)}}.$$

The electricity supplied each year can be estimated as:

$$\text{Electricity per year} = I \times A E_{\text{m}} \times E_{\text{s}} \times 365 \text{ kWh}$$

where I = average annual irradiation in kWh/m^2-day
  A = array area in m^2
  E_{m} = module efficiency
  E_{s} = system efficiency
  365 = days in a year.

When comparing two systems, it is often useful to see how the unit electricity cost varies depending on the size of the system, or to examine the effect of varying the module price, or the diesel fuel price. Graphs illustrating these variations can be computed by hand, or preferably using a spreadsheet program. An example comparing the electricity cost of PV with that of diesel or an extension of the grid is shown in Fig. 5.2.

The life-cycle costing process may appear rather long and complex, but if a little care is exercised it provides a relatively straightforward way to make a valid economic comparison between different options. Taken with other non-economic factors, this forms a vital part of the decision making process.

5.2.3 Conclusion

The calculation laid out in Table 5.5 indicates that a PV system would be marginally more expensive than the diesel alternative. However, with the margin of error of a 20 year calculation, the two systems should be regarded as similar. It must be stressed that this calculation alone should not determine which system is the better option. There are other factors to be taken into account, in particular:

- *Reliability*. Although estimated costs for operation and maintenance (O&M) and component replacement were included in Table 5.5, further consideration must be given to the overall effect of a system breakdown. What will be the consequence of losing power for a number of hours or days? Depending on the application, this might mean lost revenue, decay of refrigerable goods, a lack of water, etc. In general, PV is considerably more reliable than the alternatives and this can greatly enhance its value to the customer.

- *Social and environmental benefits.* Other disadvantages of diesel systems which are hard to quantify but make them inferior to PV, are in areas such as noise, fumes, and vulnerability to fuel shortages. Against these one must also be aware of the greater land area required by the PV system which may have to be paid for, and its dependence on weather conditions.

The graph in Fig. 5.2, computed for an insolation level of 5.5 kWh/m² day indicates that on the basis of life-cycle costs, diesel will be cheaper than PV for anything above the smallest of loads (2–3 kWh/day) at the current PV system cost of around $10/W. However, with the anticipated fall towards $3/W over the next 20–30 years, PV will become economically attractive at much greater loads.

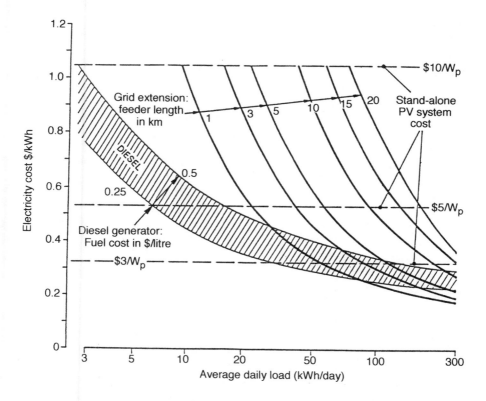

Fig. 5.2 Unit electricity cost against load. Comparison of PV, diesel and grid extension

Summary

The method of life-cycle costing is the usual method for determining the economic viability of an application. We have shown how this method can be used to calculate the annualised life-cycle costs as well as the cost of generated electricity for PV systems, and thus provide an important input to the design of the system. An economic comparison has been made between PV systems and alternative energy sources.

5.3 RURAL ELECTRIFICATION

5.3.1 Introduction

The provision of electricity to rural areas derives important social and economic benefits to remote communities throughout the world. Power supply to remote houses or villages, electrification of the health care facilities, irrigation and water supply and treatment are just a few examples of such applications.

The potential for PV-powered rural applications is enormous. The UN estimates that two million villages within 20° of the equator have neither grid electricity nor easy access to fossil fuel. It is also estimated that almost a half of all people worldwide do not have electricity, with a large number of these people living in climates ideally suited to PV applications. Even in Europe, thousands of houses in permanent occupation (and yet more holiday homes) do not have access to grid electricity.

As we have seen in section 5.2, the economics of PV systems compares favourably with the usual alternative forms of rural electricity supply, grid extension and diesel generators. The extension and subsequent maintenance of transmission lines over long distances, and often over difficult terrain, is expensive, particularly if the loads are relatively small. Regular fuel supply to diesel generators, on the other hand, often present problems in rural areas, in addition to the maintenance of the generating equipment.

5.3.2 Domestic supply

Stand-alone PV domestic supply systems are commonly encountered in developing countries and remote locations in industrialised countries. The size range varies from 50 W to 5 kW depending on the existing standard of living. Typically larger systems are used in remote locations or island communities of developed countries where household appliances include refrigerator, washing machine, television and lighting. In developing regions large systems (5 kW) are typically found for village supply while small

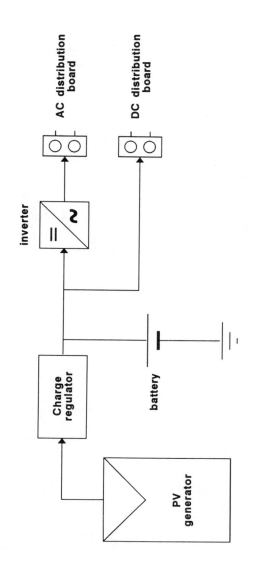

Fig. 5.3 Configuration of a stand-alone residential PV system

systems (20–200 W) are used for lighting, radio and television in individual houses.

A typical configuration of these systems is shown in Fig. 5.3. The inverter is omitted in systems where AC power is not required.

(a)

(b)

Fig. 5.4 (a) The Beginish Island PV system; (b) Rappenecker Hof

Example: Beginish Island Electrification (Fig. 5.4a). Jim and Mike Casey, two elderly brothers, are the only inhabitants of little Beginish Island off southern Ireland. Up until 1983, they rowed to the mainland every Sunday for Mass and to get the battery charged. Then the Caseys got a PV system to charge the battery which now powers the television, two 12 V, 15 W fluorescent lights and a VHF radio.

The complete PV system consists of four 19.2 W modules, two 12 V car batteries, a diode and a voltmeter. Operation and maintenance simply involve occasionally cleaning the array modules and adding distilled water to keep the battery plates covered. The Caseys control charging by manually adjusting the operating time of the television, lights and radio according to the amount of sunlight. They prevent deep discharge by turning off all the appliances when the voltmeter reading drops to 11 V.

This simple PV system has operated reliably and continuously since 1983.

Example: Rappenecker Hof, Germany (Fig. 5.4b). This 17th century farmhouse, now converted to an inn, has been powered by a PV system since 1987. When the sunlight is strong, the PV system supplies all the energy needs of the inn. A diesel generator, originally used to power the inn, backs up the PV system when the energy demand is high and when the weather is bad. The combined system is just as reliable as the utility grid. For example, in 1988 the total energy consumption at the inn was 2797 kWh. Of this, 77% was supplied by the PV system and the remainder by the diesel generator. (Recently, a 1 kW wind generator was installed to reduce the demand for diesel back-up.)

5.3.3 Health care systems

Extensive vaccination programmes are in progress throughout the developing world in the fight against common diseases. To be effective, these programmes must provide immunisation services to rural areas. All vaccines have to be kept within a strict temperature range throughout transportation and storage. The provision of refrigeration for this aim is known as the *vaccine cold chain* (Fig. 5.5).

Vaccine refrigerators are required to maintain vaccine between 0°C and 8°C at all times. In addition, a separate freezing compartment is usually needed to freeze ice packs which are used for transporting vaccines in cold boxes. Kerosene refrigerators are unreliable, and both bottled gas and diesel often suffer from fuel supply problems. Solar photovoltaic refrigerators (Figs. 5.6 and 5.7a) have shown that they are able to provide a more sustainable vaccine cold chain. In particular, they offer greater reliability, longer equipment lifetime and better temperature control. Over the past five years, over 2000 photovoltaic medical refrigerators have been installed in Africa.

Fig. 5.5 The vaccine cold chain

There are, however, also some disadvantages. The main ones are that the repair work requires a skilled technician, the planning and installation of the system takes longer, and that the local staff have to be carefully trained to use the system.

A solar photovoltaic refrigerator is likely to cost around $3000–$4000, more than a kerosene unit ($600–$800). The latter, however, will use 0.5 to 1.0 litres of fuel per day, require frequent maintenance, and have a shorter lifetime. In general, the total cost over, say, 20 years will be approximately equal for the solar and kerosene refrigerators. But, because of the greater reliability and the resultant saving in wasted vaccine, solar refrigerators are the preferred option.

5.3.4 Lighting

Lighting is taken for granted in the industrial countries and in most of the urban areas of the developing countries. In areas without access to mains electricity, lighting is restricted to candles, kerosene lamps, or torches powered by expensive throw-away batteries.

In terms of the number of installations, lighting is presently the biggest application of photovoltaics, with tens of thousands of units installed world-wide. They are mainly used to provide lighting for domestic or community buildings, such as schools or health centres. PV is also being increasingly used for lighting streets and tunnels, and for security lighting (Figs. 5.7b and 5.8).

The performance of PV lighting systems has been excellent, with increasing demand for more systems in the locality where a PV light is installed.

Solar array

Inside the
health clinic

Health
centre

Charge controller

Batteries

Fig. 5.6 Schematic diagram of PV refrigerator

The use of low-voltage DC fluorescent lamps, rather than filament lamps, is important for efficient use of the electricity. AC lights can be used with an inverter but this introduces electrical losses.

Even though individual PV lights are more expensive to buy than kerosene lamps, they are generally more cost-effective over the lifetime of the installation. In addition, they provide better quality illumination.

5.3.5 Battery charging

In remote locations, batteries are often used as the main source of electricity. Ordinary car batteries are used not only in vehicles but also to power

Fig. 5.7 (a) The vaccine refrigerator; (b) PV security lighting

Typical PV lighting system

Portable light with
integral PV cells

Area or security light

Fig. 5.8   PV lighting systems

lights, televisions and radios. The usual method for charging batteries has
been to use diesel or gasoline generators at battery-charging stations in the
nearest village, or to transport the batteries periodically to a mains supply.
PV systems are now also common, either at battery-charging stations or in
small dedicated systems (Fig. 5.9).

   A PV battery-charging system consists simply of a PV array with support
structure and charge controller. They are usually suitable for insolation
levels above 3.0 kWh/m^2 day.

Summary

Photovoltaic systems provide a natural means for rural electrification.
Various types of these systems have been described, with more detailed

A 1400 W$_p$ battery charging station

Fig. 5.9 PV battery charging

discussion devoted to domestic supply, health care systems, lighting and battery charging. Attention has been focused on the social benefits that this type of electrification brings to the rural communities.

(a)

(b)

Fig. 5.10 (a) PV pump and storage tanks; (b) PV pumping with a concentrator array (courtesy of Midway Labs Inc.)

5.4 WATER PUMPING

5.4.1 Introduction

More than 10 000 PV-powered water pumps are known to be successfully operating throughout the world. Solar pumps are used to pump from boreholes, open wells, rivers and canals to provide water for villages and irrigation supplies. Less-common applications include drainage pumping and water circulation for fish farms. The largest number of solar pumps in one country is in India where more than 500 systems have been installed for village water supplies.

A solar-powered pumping system (Fig. 5.10) consists of a PV array powering an electrical motor which operates a pump. The water is pumped up through a pipe and into a storage tank. The energy from the PV array is therefore converted into the potential energy of the pumped water. In most instances, this obviates the need for battery storage of the generated electricity.

Solar pumps are used principally for two applications: village water supply (including lifestock watering) and irrigation. These two applications have very different demand patterns: villages need a steady supply of water, whereas crops have variable water requirements during the year.

A solar pump for village water supply is shown schematically in Fig. 5.11. Note that water has to be stored for periods of low insolation. For example,

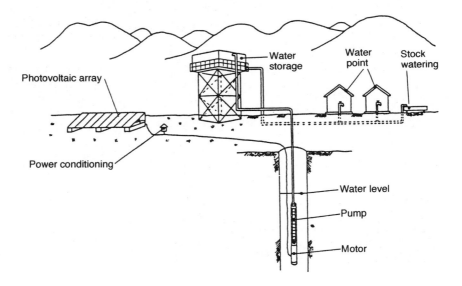

Fig. 5.11 Diagram of solar-powered village water supply

Fig. 5.12 Diagram of solar irrigation pumping system

in the Sahel region of Africa the storage needs to be 3–5 days of demand. During rainy seasons (which also coincides with the period of lowest solar radiation), the reduced output of the pump can be offset by capturing the rain water. The majority of solar pumping systems installed to date are for village water supply.

A solar pumping system for irrigation is shown in Fig. 5.12. In this application, the peak demand (during the irrigation season) is often more than twice the average demand. The solar pump is therefore under-utilised for most of the year.

5.4.2 The technology

Systems are broadly configured into five types (Fig. 5.13).

(a) *Submerged multistage centrifugal motor pumpsets* are probably the most common type of solar pump used for village water supply. The advantages of this configuration are that it is easy to install (often with lay-flat flexible pipework) and the motor pumpset is submerged away from potential damage. The most commonly employed system consists of an AC pump and inverter with a PV array of less than 1500 W, but DC motors are also used.

Fig. 5.13 PV pump configurations

(*b*) *Submerged pumps with surface mounted motor* were widely installed with turbine pumps in the Sahelian West Africa during the 1970s. This configuration easy access to the motor for brush changing and other maintenance, but, in general, it is largely being replaced by the submersible motor and pumpset.

(*c*) *Reciprocating positive displacement pumps* (often known as the jack or nodding donkey) are very suitable for high-head, low-flow applications where they are often more efficient than centrifugal pumps. Reciprocating positive displacement pumps create a cyclic load on the motor which, for efficient operation, needs to be balanced. Hence the above-ground component is often heavy and robust, and power controllers for impedance matching are often used.

(d) *Floating motor-pump sets* have a versatility that makes them ideal for irrigation pumping from canals and open wells. The pumpset is easily portable and there is a negligible chance of the pump running dry. The solar array support often incorporates a handle or 'wheel barrow' type trolley to enable transportation.

(e) *Surface suction pumpsets* are not recommended except where an operator will always be in attendance. Self-start and priming problems are often experienced in practice, and it is impractical to have suction heads of more than 8 metres.

5.4.3 Sizing and cost

Sizing of solar pumps is determined by the hydraulic energy required:

Hydraulic energy (kWh/day)
$$= \text{Volume required (m}^3\text{/day)} \times \text{head (m)} \times \text{water density} \times \text{gravity}$$
$$= 0.002\,725 \times \text{volume (m}^3\text{/day)} \times \text{head (m)}$$
Solar array power required (kW$_p$)
$$= \frac{\text{Hydraulic energy required (kWh/day)}}{\text{Average daily solar irradiation (kWh/m}^2\text{ day)} \times F \times E}$$

where F is the array mismatch factor, equal to 0.85 on average, and E is the daily subsystem efficiency, typically between 0.25 and 0.4.

A PV system pumping 25 m^3/day through a 20 m head requires a solar array of approximately 800 W$_p$ in the Sahel region of Africa. Such a pump would cost approximately $6000. Other example costs have already been considered in section 5.2. A range of prices is to be expected, since the total system comprises the cost of modules, pump, motor, pipework, wiring, control system, array support structure and packaging. Systems with larger array sizes generally have a lower cost per W. The cost of the motor pumpsets varies according to application and duties; a low-lift suction pump may cost less than $800 whereas a submersible borehole pumpset costs $1500 or more.

In general, photovoltaic pumps are economic compared to diesel pumps up to approximately 3 kW$_p$ for village water supply and to around 1 kW$_p$ for irrigation.

Summary

Solar pumping is an important application of photovoltaics which does not need battery storage. Two important applications have been discussed, village water supply and irrigation. Different types of pumps have been reviewed, alongside their suitability for different uses. The sizing and typical prices of systems have been considered.

5.5 PROFESSIONAL APPLICATIONS

5.5.1 Introduction

The installations of PV systems which are discussed in this section are fully justifiable on the grounds of conventional economic arguments which were discussed in section 5.2. These applications include telecommunication systems (radio transceivers, telephone systems, etc.), remote monitoring and

Fig. 5.14 PV-powered radio telephone system

control (scientific research, seismic recording, climate monitoring, traffic data collection, etc.), signalling systems for railways or river and ocean navigation aids, cathodic protection of remote pipelines or bridges, and many other applications. Some selected applications of this type will now be discussed in more detail.

5.5.2 Telecommunications and remote monitoring

Photovoltaic systems have had more commercial success in telecommunications and remote monitoring than in other remote power applications. They can be effective at relatively low insolation levels compared to other applications. Fig. 5.15 shows two examples of such systems: a hilltop TV translator and PV-powered weather monitoring system.

Data collection from remote monitoring stations usually takes place by UHV/VHF radio transmission, when the central computer contacts the remote system for data. As well as transmitting information (for example, gas pressure) the system can also transmit the state of charge of the battery, thereby giving the central computer operator an advanced warning of system failure due to lack of battery power.

These applications need small but reliable power supplies which can be easily provided by a PV system charging a battery. Photovoltaics is ideal because the system can be sized to provide the small amounts of energy required for transmission and reception while requiring little or no maintenance.

The economics of PV are here very competitive with other forms of power supply. Mains power is unlikely to be available in the areas where remote monitoring systems are installed. Even if a mains cable were nearby, extending it more than a metre to supply a small amount of power is normally uneconomic.

Primary batteries are still the most commonly seen power source for remote systems. Initial cost comparisons with PV look attractive only until the long-term costs such as regular battery replacement, and the related personnel and vehicle costs, are accounted for.

To operate unattended for long periods, various hazards must be avoided which include theft or vandalism of the system, damage by animals or falling branches, shading by growing vegetation, and snow covering.

Since the reliability of these systems is paramount, the sizing method used is based on radiation data for the worst month of the year (usually December in the northern hemisphere) rather than on the average daily irradiation over the year. In addition, to ensure system operation even over periods of lower solar radiation than predicted from the average yearly values, the sizing often uses data for the worst year in ten. An example of system sizes that have been used in practice are given in Table 5.6.

Fig. 5.15 (a) PV-powered weather monitoring: (b) Hilltop PV translator

Example (British Gas plc, UK). The system shown in Fig. 5.16a powers a UHF/VHF transmitter via a 12 V sealed lead-acid battery. The sizing of this system has been carried out according to the data in Table 5.6.

Table 5.6 Example of array and storage sizes for remote monitoring applications in Nothern Europe, determined from the December Solar radiation data for an average year and for the worst year in ten (values given produce 1 Wh/day)

	Average	Worst
Array size (W)	13.5	110
Battery size (Wh)	50	165

Example (Guide Dogs for the Blind Association, UK). The GDBA orientation-assisting device consists of a series of beacons (as shown in Fig. 5.16b) strategically positioned along specific routes which transmit preselected location radio codes to a chosen distance. A receiver carried by a visually handicapped person responds to the beacon radio codes by playing a prerecorded audible message, giving orientation and directional information (where the shops are, road crossings, etc.).

5.5.3 Cathodic protection

Cathodic protection is a method for shielding metalwork from corrosion, for example, pipelines and other metal structures. A PV system is well suited to this application for two reasons:

- Power is often required in remote locations along the path of a pipeline.
- Cathodic protection requires a DC source of electricity.

An example of such a system in shown in Fig. 5.16c.

Corrosion is a chemical reaction between metal, usually iron, and water, which turns the iron into iron oxide (i.e., rust). You will remember from your chemistry lessons that oxidation involves electron transfer to the metal. If an electricity source is connected to the metalwork which sets it at a negative voltage (thus making it a 'cathode'), the iron is prevented from oxidising. The positive terminal of the source must be connected to an anode, often a piece of scrap metal, which will corrode instead.

Even if cathodic protection is used, the pipeline or structure must still be protected in the normal way with weather-resistant paint. Otherwise the PV array would have to be very large to cope with the high rate of corrosion.

(a)

(b)

(c)

Fig. 5.16 (a) PV-powered remote monitoring; (b) OAD Beacon; (c) Cathodic protection of a gas well

Summary

A range of commercially viable applications of PV systems has been discussed. These systems usually operate unattended and often in isolated locations. The need for reliability in sizing has been emphasised.

5.6 ELECTRIC POWER GENERATION IN SPACE

5.6.1 Introduction

Photovoltaic solar generators have been and will remain the best choice for providing electrical power to satellites in an orbit around the Earth. Indeed, the use of solar cells on the US satellite Vanguard 1 in 1958 demonstrated beyond doubt the first practical application of photovoltaics. Since then, the satellite power requirements have evolved from a few watts to several kilowatts (Figs. 5.17 and 5.18) with arrays approaching 100 kW being planned for launch within the next decade (Fig. 5.19).

A space solar array must meet strict criteria, and the most important considerations can be summarised as follows:

- Extremely high reliability in the adverse conditions of the space environment.
- High power/weight ratio.
- Cost constraints on the PV system are of secondary importance.

The satellite applications which are driving the European space photovoltaic technology are illustrated in Fig. 5.20. Eureca (European Retrievable Carrier) is the first European spacecraft designed to return to ground after operating for just a year in low-earth orbit (LEO, with altitude usually less than 1000 km). Olympus is a large telecommunications satellite designed for a planned ten-year lifetime in geostationary orbit (GEO, at altitude of about 36 000 km). Agora (Asteroid-Gravity Optical and Radar Analyser) is a scientific mission which aims to explore the asteroid belt at a distance of 2.5 astronomical units from the Sun. Columbus is Europe's space station related concept which is currently in preparatory programme phase.

5.6.2 Satellite PV system

In common with their terrestrial counterparts, the satellite PV system is usually divided into the photovoltaic generator, the storage subsystems and power conditioning which includes also distribution.

EURECA SOLAR ARRAY

- **FEATURES:**
- RARA-Technology
  (Retractable Advanced
  Rigid Array)
- Re-usable (5 flights)
- Bus Voltage: 28+/-4Volts

- **CHARACTERISTICS:**
- 5 kW (EOL/ 1yr)
- 5 Panels per Wing
- Panel Size: 3.4m x 1.4m
- Mass per Wing: 120 kg

esa

Fig. 5.17   The Eureca PV array (courtesy of the European Space Agency)

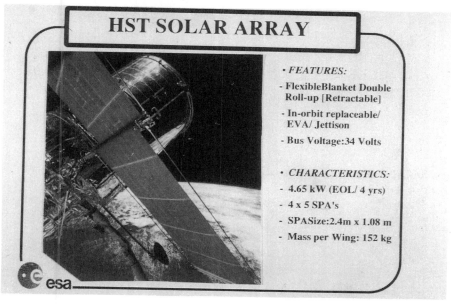

Fig. 5.18   The Hubble space telescope array (courtesy of the European Space Agency)

Space Platform Characteristics

	Re Carrier	Earth Obs.	Power Mod.	Telecom
1st Generation Array Power [KW-BOL]	6	2	20 − 30	10
2nd Generation Array Power [KW-BOL]	10 − 20	8 − 16	30 − 100	10 − 30
Lifetime [Yrs]	5 × 0.5	5	10	10
Bus Voltage [V]	28	28	120	50
Partial Deployment	−	−	+	+
Retraction	+	−	+	−

Fig. 5.19 The space station (courtesy of the European Space Agency)

Spacecraft Characteristics

	Eureca	Agora	Columbus	Telecom
1st Generation Array Power [KW-BOL]	6	24	30	10
Lifetime [Yrs]	5 × 0.7	5		10
Bus Voltage [V]	30	50	120	50
Partial Deployment	–	–	+	+
Retraction	+	–	+	–
2nd Generation Array Power [KW-BOL]	10–20	–	> 100	10–30

Fig. 5.20 European space projects driving the solar array technology (courtesy of the European Space Agency)

Most satellites require energy storage for two reasons:

- During the eclipse period, the solar array provides no power and at least certain vital equipment has to be supplied with electricity from batteries which are recharged in the sunlit part of the orbit. In some cases, the full eclipse power for the payload is required.
- For peak power requirements which exceed the solar array capacity, batteries are required to act as an energy buffer.

The output power from the satellite's PV generator changes with the orbital position of the satellite, the season, and the age of the system. The power-conditioning subsystem provides the necessary regulation before the power is distributed to the various users, and also controls the charging and discharging of batteries.

5.6.3 The PV generator

The most obvious requirement on the design of a satellite solar array is its absolute power level, resulting from the specific mission requirements. The power-to-mass ratio is another important factor reflecting the importance of a lightweight design dictated by the very high cost of launching a kilogram of spacecraft into orbit. The evolution of solar array technology with respect to these two parameters is shown in Fig. 5.21. Of high importance in the array design are also the environmental conditions which vary considerably from mission to mission, and are discussed in more detail below.

The space solar array must generally have good resistance to the particle and UV irradiation. The particle flux is particularly damaging in orbits which pass through the van Allen belts of electrons and protons trapped by the electromagnetic field of the Earth.

Satellites in a low-earth orbit move in a relatively benign radiation environment but face other hazards due to residual atmosphere. The ionised plasma causes power loss in high-voltage arrays due to leakage, as well as problems associated with electric discharge. The presence of atomic oxygen causes erosion of exposed surfaces, and large solar arrays experience air drag.

Another aspect of the orbital missions, particularly important for arrays designed for long life, are the sun/eclipse thermal cycles which may cause fatigue of solar cell interconnections

The space solar cells, therefore, must have high operation efficiency, high resistance to irradiation by energetic particles (mainly electrons and protons) and a good power-to-weight ratio. The historical development of these cells is summarised in Fig. 5.22.

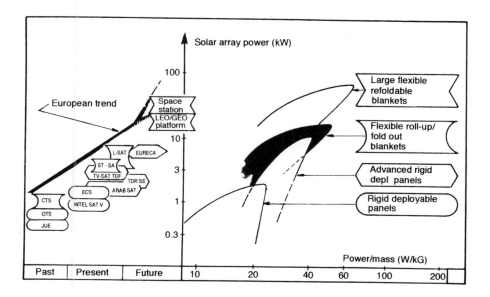

Fig. 5.21 Evolution of space solar arrays

Following the early Si cells based on n-type substrates and p-type diffused emitters, the cells in use since the early 1960s have been based on p-type substrates because of their higher radiation resistance. Subsequent developments include the incorporation of the back-surface field (BSF), back-surface reflector (BSR), the use of a shallow emitter (the 'violet cell'), and top surface texturing (the 'black cell').

Special monocrystalline silicon solar cells have been used almost exclusively up till the present, principally because of the well-tried technology. Gallium arsenide cells, which have higher efficiencies and higher radiation tolerance, have been under consideration (but only in limited use) since

Fig. 5.22 Historical development of space solar cells

the 1970s. Over the last decade, active research into indium phosphide cells has been gaining momentum because of their high radiation resistance. As with terrestrial cells, the search for new materials and structures is continuing.

Summary

Photovoltaics has been the principal means of electricity supply to satellites since the birth of the space era. Various satellite power systems have been examined. The criteria placed on space PV arrays have been discussed in detail, alongside the structure of the complete system. A brief historical review of space solar cells has been presented.

5.7 GRID-CONNECTED SYSTEMS

5.7.1 Introduction

Grid-connected* systems are now being seriously considered to supplement the conventional power generation in many industrialised countries. Indeed, there are already many grid-connected demonstration projects which study this possibility. Although they have yet to become viable on economic grounds, the participation of PV in large-scale power generation is viewed with increasing prominence as a means of halting the adverse environmental effects of conventional energy sources (see Chapter 6).

Two types of grid-connected installations will be reviewed in this section: centralised large-scale PV power generation in PV power stations, and an alternative, distributed form of power generation in units located directly at the customer's premises.

5.7.2 PV power stations

A PV power plant feeds the generated power instantaneously into the grid by means of one or more inverters and transformers. Most larger systems use line-commutated inverters equipped with a maximum-power-point tracker (Fig. 5.23a). Fig. 23b shows the first PV power station, built at Hysperia in southern California in 1982 with nominal power specification 1 MW. It uses crystalline silicon modules mounted on a two-axis tracking system.

---

* The 'grid' is used here as a generic term for the electricity distribution network. In some countries, notably in the U.K., the 'grid' refers to the high-voltage transmission system to which PV systems are unlikely to be connected! I am grateful to Professor R. J. Arnold of Southampton University for pointing this out.

Fig. 5.23 (a) Configuration of PV power plant with one or several inverters; (b) PV power station (courtesy of P. D. Maycock, Energy Systems Inc.)

There is little doubt that, at present, the cost of generated electricity from PV plants is not generally competitive with the conventional power sources: the cost of electricity from PV power plant is typically 10–15 times higher than from a large thermal plant. Additional consideration must also be given to the area of the system as, for example, a 25 kW plant represents an active PV array surface of approximately 250 m², and occupies a land area of roughly 1000 m².

There are, however, some exceptions. It may be already economic to set up a PV plant to assist the local grid during periods of peak demand rather than construct a new power station. This is known as *peak shaving*. It can also be cheaper to place a small PV plant at the end of a transmission line rather than upgrade the line so that it can take more power from the grid. This application is known as *embedded generation*.

PV power plants are discussed further in Chapter 7 where we consider large-scale PV projects.

5.7.3 PV in buildings

5.7.3.1 Introduction

PV arrays mounted on roof tops and facades offer the possibility of large-scale power generation in decentralised medium-sized grid-connected units. The PV system supplies the electricity needs of the building, feeds surplus electricity into the grid to earn revenue, and draws electricity from the grid at low insolaton (Fig. 5.24). This potential type of PV operation within the power generation sector is presently favoured in many industrialised countries. The significant potential for this technology can be seen from the following example.

Fig. 5.24 Roof-top grid-connected PV system

Example. It has been estimated (Bloss *et al.*, 1991) that in Germany there are 650 km^2 of roof area which is technically suitable for PV installations. Assuming 8% system conversion efficiency, this represents a distributed power plant of 52 GW$_p$. This value should be still higher if the potential of facades were included.

A large number of roof-mounted PV systems are now being installed in Germany within the government sponsored 'Thousand Roofs' programme. The size envisaged for the decentralised residential PV systems in typically between 1 and 5 kW$_p$.

The main advantages of these distributed systems over large PV plants are as follows:

- There are no costs in buying the land and preparing the site,

- The transmission losses are much lower because the load is on the same site as the supply.

- The value of the PV electricity is also higher because it is equal to the *selling price* of the grid electricity which has been replaced, rather that to the cost of generating it.

However, it should also be noted that the price paid by utility companies for electricity exported from a decentralised source is not the same as the utility sale price of electricity. In some countries, utilities do buy PV electricity at a similar price to which it is sold. In other countries, however, PV electricity is bought at a considerably lower rate. The optimum economic benefit is then derived by consuming all PV-produced electricity, with direct reduction of the energy imported from the utility. Thus grid-connected PV systems are ideal for loads which vary in proportion to the irradiation. Typical loads are air-conditioning, refrigeration and pumping. Other significant loads can be timed to operate when PV power is likely to be available. Examples include washing machines and clothes dryers which can operate on timing clocks.

5.7.3.2 The PV system

The PV system diagram is shown in Fig. 5.25. It consists of a photovoltaic generator connected via a DC protection and isolation unit to the inverter. The inverter supplies power to the domestic distribution board, or the surplus to the utility grid, via two meters which measure the input and output power of the PV system. The power fed to the grid must be of sufficient quality which is usually determined by the inverter characteristics. In particular, the total harmonic distortion and the number

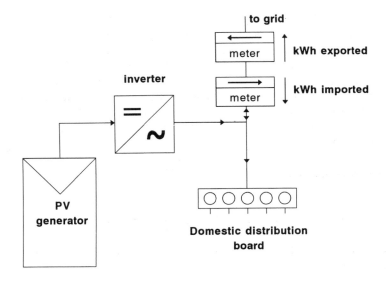

Fig. 5.25 Configuration of a residential grid-connected system

of harmonics must be low. It is often desirable to maximise the power transfer from the PV generator by employing the maximum-power-point tracker.

The best price of a grid-connected PV system integrated into a roof structure in 1990 was estimated to be about 7.4 $/W, including the modules (5.5 $/W), inverter (1.25 $/W) and installation (0.75 $/W).

Example. The Sprayfield domestic PV system is a 5.5 kW$_p$ photovoltaic system integrated into a domestic dwelling near Cork, Ireland. The main aim of the project was to investigate the feasibility of integrating a photovoltaic system into a typically Irish dwelling house, the feasibility of using a PV system to supply the energy demand of a dwelling in a Northern European latitude, and to demonstrate the reliability of the gird-connected PV system. The project was a part of a CEC DGXVII demonstration programme.

The PV generator has been installed on a metal structure at a tilt angle of 30°. The power-conditioning unit is situated in an enclosed cabinet located in the garage. The data monitoring system was also located in the garage. The siting and the panel inclination were selected to provide the most cost-effective solution (a computer model showed that any angle between 20° and 45° would be suitable).

The inverter supplies energy to the utility grid or to all loads in the house, namely washing machine, dryer, dishwasher, cooker and lights.

The photovoltaic generator consists of 144 modules, and the inverter is a line-commutated single phase 7 kVA inverter. It is controlled by a maximum-power-point tracker using the constant-voltage temperature-compensated method. A small weather station records the meteorological data (temperature and irradiation). The array DC energy and voltage, and the AC energy to loads and to the grid are also recorded. Data monitoring is also carried out by a dedicated computer with data being transferred by modem to a remote personal computer.

The system performance is shown in Fig. 5.26 which shows the PV power supplied to the load, exported or lost throughout the year. Fig. 5.27 shows the energy supply to the load by PV and the grid.

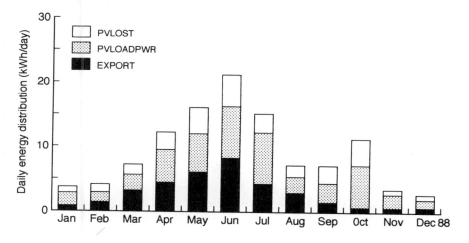

Fig. 5.26 System performance of the Sprayfield system: daily energy distribution

From a technical point of view the installation was a success as this project has shown that a grid-connected PV system is technically viable in an Irish location. Economically, however, the cost of PV-generated electricity is presently not competitive with the price of electricity received from the grid.

Summary

Photovoltaics is now being seriously considered as a potential large-scale source of electric power in grid-connected applications. Two principal types of such systems have been discussed: PV power stations and a decentralised range of PV systems in buildings. The latter form has been highlighted by

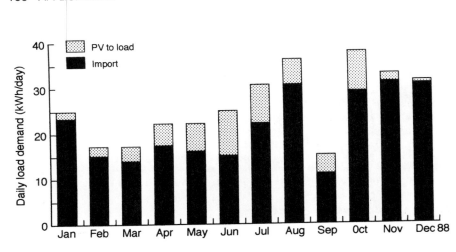

Fig. 5.27 System performance of the Sprayfield system: energy supplied to the load

giving a decription of the potential scale of applications, the system structure, and its interaction with the utility.

SUMMARY OF THE CHAPTER

A range of photovoltaic applications has been examined. Most of these were stand-alone small- to medium-sized systems which exploit the reliability of PV power supply and require little maintenance, making them suitable for remote or isolated locations.

Economic analysis based on life-cycle costing has been analysed in detail and applied to PV systems. A comparison with alternative electricity sources in rural locations has been made, with the conclusion that PV systems provide a good option in a range of applications which require grid extension or small diesel generating systems.

Rural PV electrification has been discussed with emphasis on the social benefits that this brings to rural populations. Particular attention has been given to solar pumping where PV systems have been applied with success for some time.

Professional PV applications have been described, and their special features highlighted. Space PV systems, an important application of PV systems since the early days of modern photovoltaics, have been discussed in some detail.

In conclusion of the chapter, an overview has been given of potential grid-connected systems that are under consideration for large-scale power generation. Both centralised and decentralised types of such systems have been discussed.

BIBLIOGRAPHY AND REFERENCES

BOGUS, K. Space photovoltaics—present and future, *ESA Bulletin* 41, 70–77.

HILL, R., *Applications of Photovoltaics*, Adam Hilger Bristol, 1988.

MARKVART, T. Radiation damage in solar cells, *Journal of Materials Science: Materials in Electronics* **1** (1), 1990: 1–8.

MAYCOCK, P. D. and STIREWALT, E. N., *A Guide to the Photovoltaic Revolution*, Emaus, PA., 1985.

BLOSS, W. H., PFISTERER, F., KLEINKAUF, W., LANDAU, M., WEBER, H. and HULLMAW, H. Grid-connected solar houses, In: *Proc. 10th European Photovoltaic Solar Energy Conf., Lisbon*, 1991: 1295–1300.

SELF-ASSESSMENT QUESTIONS

PART A. True or false?

1 PV can already be an economically viable source of electricity for rural electrification.

2 The annualised life cycle cost of a PV system is the total life-cycle cost divided by the number of years when the system will remain in operation.

3 PV-powered remote monitoring systems can only be installed in locations with high solar radiation.

4 Domestic PV supply is usually designed for lower reliability than PV-powered telecommunication systems.

5 Multicrystalline silicon cells are a frequent choice for space applications because of their low cost.

6 A major advantage of PV systems is their reliability.

7 A large component of the cost of a PV pumping system is the battery cost.

8 PV can potentially provide a considerable part of the household electricity supply in many industrialised countries.

9 PV provides a good source of electricity for water heating.

10 The largest PV power station in the world has 150 MW of installed power.

PART B

1 Name some applications where the installation of a PV system is fully justifiable on the grounds of conventional economic arguments.

2 What advantages do PV systems have in comparison to diesel pumping?

3 Outline the demand characteristics of: (a) Village water supply; (b) Irrigation pumping.

4 What are the principal requirements on space solar cells? What are the principal hazards that affect the satellite array design?

5 What are the main advantages of distributed PV systems over centralised power stations? What is the general configuration of the system?

PART C

1 You are setting up a PV-powered battery supply. You have the choice between a battery costing $500 with a life of 5 years, or a battery costing $1100 guaranteed to last 15 years. What are the life-cycle costs for each option over 15 years if the discount rate is 10%? Would your choice change if the discount rate was only 5%?

2 A cheap diesel generator costs $500 and has a fuel bill of $100/year. A more efficient generator would cost $800 and cut the fuel bill to $75/year. Both will last 10 years. Which generator would work out more cost-effective at a discount rate of 10% and an inflation rate for diesel of 5%? What is the present worth of all the fuel consumed by each generator?

3 Use the layout of Table 5.4 and a discount rate of 10% to calculate the annualised life-cycle cost of the following system: a 1000 W_p PV array costing $8/$W_p$ and likely to last for 30 years, two batteries costing $400 each that need replacing every 5 years, a power control unit with a 10-year lifetime costing $800, and $50 of wiring that won't need replacing. Assume that you will install the system yourself but that the annual operation and maintenance costs will be 1% of the initial capital cost.

4 If the battery and control unit in the preceding question are 80% efficient, and the array consists of modules of 14% efficiency, what will be the unit electricity cost of the system if it is situated in a region with 5.5 kWh/m^2 day average daily irradiation?

5 A village has a population of 350 people who require clean water, and intend to install a PV pump. Assume a demand of 40 litres per person per day, and that the static water table is 15 m below the surface.
   (a) What volume of water per day (in m^3) is required?
   (b) For two days of storage, what volume (in m^3) of tank would be needed?
   (c) What is the total head over which water must be pumped? Assume a drawdown of 5 m once pumping begins and that the top of the tank will be 3 m above ground level.
   (d) What is the daily hydraulic energy required in kWh/day?
   (e) What is the size of PV array needed, in W_p? Assume a daily solar irradiation of 5.0 kWh/m^2 day, an array mismatch of 0.8, and a daily subsystem efficiency of 0.3.

Answers

PART A

1, True; 2, False; 3, False; 4, True; 5, False; 6, True; 7, False; 8, True; 9, False; 10, False.

PART B

1 For example, remote telecommunication systems, remote monitoring and control systems (seismic recording, climate monitoring, traffic data collection, etc.), signalling systems or cathodic protection.

2 There are no fuel costs or fuel supply problems; PV systems can usually operate unattended, are very reliable, and require little maintenance.

3 (a) Village water demand is fairly constant throughout the year. (b) For irrigation supply, the demand will be very high in some months and almost nothing in others.

4 High efficiency combined with high resistance to particle radiation and high power-to-weight ratio. The principal hazards that affect satellite solar array design are particle and UV radiation, residual atmosphere and ionised plasma in low-earth orbit, and thermal cycles.

5 There is no cost of buying the land. Value of electricity is higher because it is generated where it is required, and transmission losses are much lower. The configuration is shown in Fig. 5.25.

PART C

1 $1005 and $1100. Yes.

2 The cheaper generator. $781 and $586.

3 LCC = $12160, ALCC = $1289/year.

4 $0.80/kWh

5 (a) 350 people × 40 l/person/day = 14,000 l/day = 14 m^3/day.
   (b) 2 days × 14 m^3/day = 28 m^3.
   (c) Total head = static level + drawdown + tank height.
              = 17 m + 5 m + 3 m = 25 m.
   (d) Hydraulic energy = 0.0027 × volume × head
              = 0.0027 × 14 m^3/day × 25 m = 0.95 kWh/day
   (e) Array size (kW_p) = Hydraulic energy / (irradiation × F × E) where F = array mismatch factor and E = subsystem efficiency.
       So array size (kW_p) = 0.95 kWh/day/(5.0 kWh/m^2/day × 0.8 × 0.3).
              = 0.79 kW_p = 790 W_p.

6

Environmental Impacts of Photovoltaics

AIMS

The aims of this chapter are to examine the environmental impacts associated with the production, use and disposal of PV modules, and the methodologies employed in the environmental assessment of PV relative to other forms of electricity generation technologies.

OBJECTIVES

On completion of this unit, the student should be able to:

1. Understand the methodology employed in the environmental assessment.
2. Assess the environmental impacts of producing PV modules using cells of silicon, copper indium diselenide and cadmium telluride.
3. Analyse the hazards associated with the use of PV modules.
4. Recommend means of final disposal or recycling of PV modules.

6.1 INTRODUCTION

The environmental damage by energy production has been known for some time, and the assessment of this damage is becoming an issue of ever increasing importance. It is therefore essential that environmental aspects of photovoltaic power generation are considered in some detail.

Written by R. Hill.

The plan of this chapter is as follows. In section 6.2, the requirements on the methodology for assessment are defined, and the concept of external costs is introduced. Section 6.3 discusses the contribution to the external costs in terms of fuel cycles for energy technologies. The environmental impacts of photovoltaic power generation are analysed in section 6.4.

6.2 THE NEED FOR ENVIRONMENTAL ASSESSMENT OF ENERGY SOURCES

It has long been recognised that energy sources are significant contributors to environmental damages. Costs associated with these damages, however, have been treated in general as external to the energy economy since the environment was regarded not as a marketable commodity but as one freely available to everyone. The costs for repairing or reversing environmental damages caused by energy sources were expected to be paid by society at large, e.g. in form of the taxation or via a reduction in the quality of life. By ignoring these costs of energy generation the energy sector based its resource acquisition decisions on information strongly biased against technologies such as renewable sources which impose smaller environmental and social costs than fossil and nuclear sources.

Only by defining and calculating the external costs of all of the different energy technologies is it possible to identify the real costs (internal plus external) associated with the generation of energy. In order to determine these costs a methodological framework has to be developed within which the external costs can be quantified. The methodology has to be simple, easily understandable and applicable to all energy technologies to allow for a comparison of different energy systems and to make sure that all major environmental impacts are internalised into the energy prices on an even footing. The methodology provides the framework within which are outlined the different steps that have to be undertaken to determine how the external costs should be derived, measured and evaluated. In a first step a definition of external costs is given, followed by the identification of the participating energy technologies and their complete fuel cycles, including extraction, processing, conversion, distribution, utilisation and final disposal. Having defined the *fuel cycle* for existing and future technologies their output of pollutants in the environment can be calculated or at least estimated. Based on these pollution outputs the impacts on the environment can be estimated and the associated costs can be derived by various economic valuation techniques. Existing computer models which have attempted to compile process data on the different steps of fuel cycles can be used to facilitate the calculation of the environmental externalities.

Endowed with this information on the total societal costs of each type of energy source, utility regulators and decision-makers have the choice to opt for the optimum energy technology for society and the environment.

Whilst, in theory, this approach seems to work well, in practice it will probably be some time before the incorporation of the state of the environment in economic terms is a realistic option. The attribution of costs to the environmental impacts of energy production and use is still in its developing stages and there are still several externalities which cannot be expressed in monetary, but only in qualitative terms (Fritsche, 1991). However, this fact should not prevent external costs being internalised in resource planning, acquisition and operating decisions as an internalisation will contribute to an improvement in energy decisions generally by sending the appropriate signals to both planners and consumers on how to satisfy the energy needs at the lowest overall costs to society (Bernow et al., 1991). There are strong indications from the recent Geneva Prepcom meetings for UNCED that environmental costs will become an internationally accepted component of traded commodities, including energy.

Summary

The concept of external costs of electricity generation has been defined, and the requirements of a methodology capable of calculating these costs have been outlined.

6.3 METHODOLOGICAL FRAMEWORK

6.3.1 Introduction

In order to quantify external costs of energy technologies a methodological framework must be developed within which the principle issues encountered in performing such an analysis can be addressed. The most significant and decisive part of any external cost analysis is to define, in as detailed a manner as possible, the full set of burdens, impacts and costs associated with different energy technologies and their fuel cycles. A number of criteria have to be considered to provide the methodology with general validity. The methodology should be:

- Simple and easily understandable, to allow laymen to make use of the methodological approach.
- Adjustable/flexible to allow for modifications and for incorporation of new items.

• Applicable to all of the different energy technologies to allow for comparisons of different energy sources on an even footing and to make sure that utility regulators and decision-makers use technologies that are environmentally benign and thus pose the lowest overall costs to society—consistent in its use of evaluation methods of environmental impacts.

6.3.2 External costs

External costs are generally divided into environmental and non-environmental costs. Most of the external costs are grouped under the heading of environmental costs, because they include some of the most significant components of external costs—the valuation of human health and life, acid deposition, greenhouse gases and global warming, to name but the most important ones. There is still a debate about non-environmental costs, about the importance of these costs, their magnitude and the extent to which they ought or ought not to be included in external cost calculation. However, if non-environmental costs are divided into the following four major types: aspect of security, natural resource management, employment and politico-economic instruments; then it is obvious that there are quite considerable societal costs involved and this indicates that non-environmental externalities should not be neglected when calculating the full costs of energy technologies (OECD, 1989).

In order to quantify environmental externalities and link them to the internal costs to derive the total societal costs of these energy systems it is necessary to find a single nominator for all aspects. Monetisation of environmental externalities is one of several valuation approaches that is aimed exactly at this: to express environmental costs in the same monetary unit as used for calculating internal costs of energy technologies (Fritsche, 1991).

Environmental impacts and the associated external costs can be grouped into four categories: negligible costs, small costs, significant costs and large costs. At each stage of the fuel cycle analysis an estimate of the physical magnitude of the environmental impacts can be given by using these four categories. Most notable among external costs are risks of damages to the environment and human health resulting from air pollutants emitted by fossil-fuel-fired generation and from the radiation risks of nuclear plant operations. Environmental costs of air pollutants in general are associated with large cost. The 'greenhouse effect', caused principally by CO_2 emissions from the burning of fossil fuels, and the resulting global warming are only just being assessed but are associated with large cost, even if there are still considerable uncertainties about the order of magnitude of the effects and the costs involved (EC, 1990).

A number of other environmental impacts, such as acid rain, risk to human health and life, and catastrophic events, will incur large costs as well. Impacts such as visual intrusion or operation accidents at the extraction stage or radiation risks may result in quite significant costs, whilst other environmental impacts such as noise or health and safety costs associated with materials inputs and construction can be expected to be small or even negligible (OECD, 1988), provided that care is taken in the design and operation of the industrial processes.

Detailed and reliable information and final figures for the environmental impacts will only become available when the valuation of these impacts has been undertaken.

6.2.3 Fuel cycles, environmental burdens and impacts

A detailed analysis of the fuel cycles of different energy systems is needed to determine the external costs incurred during the various stages of the fuel cycle process and hence to internalise these externalities into the full cost analysis of each of the energy technologies. This in turn facilitates a one-to-one comparison of different competing technologies on an even footing. Based on these data, informed policy actions/choices, either encouraging or limiting particular energy options, can be taken.

Fig. 6.1 Linkage between fuel cycle and external costs

The analysis of the fuel cycles of energy systems covers all stages of the energy production process including extraction, preparation, transport, conversion, operation, distribution, utilisation, waste processing and disposal. Each of these stages can impose environmental burdens such as, for example, acid deposition, and these burdens result in environmental impacts, such as forest damage, which in turn pose considerable costs to society (OECD, 1985). The linkages between a particular energy technology and the associated external costs are shown in Fig. 6.1.

Some of the environmental burdens and impacts are well documented and can be identified and measured. In many other cases, however, such as global warming or loss of species, the burden/impact relationships are fraught with considerable uncertainties and non-linearities. These may result from scientific uncertainties, unavailability of accurate data, spatial and temporal variabilities, or the fact that many impacts are the synergistic action between two or more burdens, e.g. the formation of ozone through the photochemical reactions between NO_x and hydrocarbons. Further work will be required to find appropriate methods of defining the burden/impact relationships which cannot be accurately measured so far.

Summary

The methodological framework for calculating external costs in monetary terms has been considered in detail. The calculation of these costs has been linked to the fuel cycles of the power generation technologies.

6.4 ENVIRONMENTAL COSTS OF PHOTOVOLTAICS

The commercial standard PV module uses solar cells made from wafers of silicon, usually 0.3 mm thick and 10 cm × 10 cm in area. A simplified 'fuel cycle' of silicon solar cells from 'cradle to grave' is shown in Fig. 6.2. The ingots may be either Czochralski-grown single crystals or directionally-solidified multicrystalline cubes. For the latter the third and fourth steps of Fig. 6.2 may be replaced by less costly steps yielding 'solar-grade' silicon rather than the semiconductor-grade material.

Each of the steps shown in Fig. 6.2 has inputs of energy and materials and requires capital equipment, and each also has potential hazards associated with it. The first step is a standard mining operation with associated hazards to the miners and inputs of diesel fuel and machinery. Metallurgical-grade silicon is made in large quantities for the steel industry, with a small fraction going as input to the semiconductor industry. The major emission of this manufacture is of silica dust which can cause lung disease, and there is a substantial energy input. The purification of silicon can involve

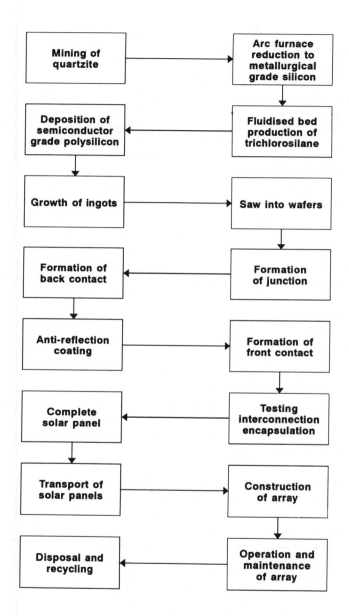

Fig. 6.2 Silicon fuel cycle

hazardous materials such as silane, whilst the doping of the silicon involves toxic chemicals such as diborane and phosphine, although only
in small quantities diluted in inert gas. These materials are used in the

microelectronics industry and their monitoring and control is well established. The materials for construction of the complete PV system, other than the PV modules, are steel, aluminium, copper, concrete and electronic equipment, with which are associated the standard industrial hazards.

The energy used in manufacturing the PV modules and the other components of the PV system is derived from the fuel mix of society and is therefore associated with emissions of greenhouse gases and acidic gases. The energy content of PV modules using silicon wafers has been measured at the Photowatt plant by Palz and Zibetta (1991) as 235 kWh(e)/m^2 for 1990 technology at 1.5 MW$_p$ per annum production rate. Thus for the 1990 crystalline silicon PV technology in small-scale production, the CO_2 emission is around 400 000 tonnes per gigawatt-year of energy output. This compares with the CO_2 output from the most modern and efficient coal-fired plant of 9 million tonnes per gigawatt-year (1991). Boiling Water Reactors have been estimated to emit about 75,000 tonnes of CO_2 per gigawatt-year (Palz and Zibetta, 1991), mainly from the fuel production steps, but this estimate did not include the energy used for decommissioning and waste treatment and storage. The CO_2 emissions estimated for the various current PV cell technologies are shown in Table 6.1, along with an estimate of values which might be typical of 2020 PV technology. The thin-film polycrystalline materials have been studied by Hynes et al. (1991) and those data in Table 6.1 are taken from their work.

As shown in Table 6.1, there are a number of technologies for PV cells other than those using wafers of silicon. Amorphous silicon cells are widely used in solar calculators, clocks, etc., but seem unlikely to be widely used for power modules. Thin multicrystalline films of silicon on ceramic substrates and thin polycrystalline films of copper indium diselenide (CIS) and of cadmium telluride (CdTe) are the three most promising technologies for efficient low-cost modules and all seem likely to be in commercial production within the next 2–4 years. An excellent review of the scientific, technical, economic and environmental issues associated with these three materials has been given by Zweibel (1990). Fig 6.3 shows a fuel cycle for CIS modules from Hynes et al. (1991).

The environmental impacts of the thin-film silicon cell are similar to those of the wafer silicon cell, but reduced in magnitude because of the smaller volume of silicon used. The CIS cell has a potential hazard in the hydrogen selenide used in its manufacture. A report on CIS manufacture by Moskowitz et al. (1990) has concluded that hydrogen selenide can be used safely provided that adequate safety precautions are adopted. There are proposals for the production of CIS which involve the use of solid selenium rather than hydrogen selenide (Badawi et al., 1991) which would largely obviate this hazard.

Both CIS and CdTe cells have a window layer of cadmium sulphide, so both types of cell could potentially present a hazard from cadmium.

Cadmium hazards arise in the refining stage, with emissions of cadmium oxide dust, emissions of cadmium during manufacture, or during a fire in which cadmium-containing PV modules were involved, and from possible leaching of cadmium from modules discarded at the end of their working life. In all of these cases, the control strategies are well established in industry, and the magnitude of the hazard is in proportion to the amount of cadmium present in the cells.

Table 6.1 Carbon dioxide emitted in the production of PV modules (in units of kilotonnes per gigawatt-year)

Cell material	Production scale	Efficiency (%)	Lifetime (years)	CO_2 kt GW^{-1} yr^{-1}
Monocrystalline silicon	small	12	20	400
	large	16	30	150
Multicrystalline silicon	small	10	20	400
	large	15	30	100
Thin-film silicon	small	10	20	130
	large	15	30	50
Thin-film polycrystalline materials	small	10	20	100
	large	14	30	40
Future (2020) multijunctions	large	30	30	24

For comparison, the most efficient coal plant emits 9 million tonnes of CO_2 per gigawatt-year, whilst a BWR nuclear plant emits 75 000 tonnes of CO_2 per gigawatt-year (plus an unknown amount for decommissioning and waste treatment).

In CIS modules, the amount of cadmium used is about 0.04 g/m^2, equivalent to 400 g per MW_p of output. This tiny amount of material is easily controlled during manufacture, whilst cadmium evaporated from modules on a building which caught fire would pose a negligible hazard to anyone far enough from the fire to avoid being burnt to death. Careless disposal of large numbers of CIS modules in one location could pose a hazard from leaching of cadmium and selenium into ground water. However, the rather scarce indium and regulations on the disposal of cadmium and selenium products would lead to recycling of the modules and reuse of the indium, cadmium and selenium.

The PV modules using CdTe cells would contain much more cadmium (about 5 g/m^2) than those using CIS cells, and the potential hazards are correspondingly higher. It is clearly beneficial to use a manufacturing process with a high utilisation of cadmium feedstock, and the success of electrochemical deposition as a production technology gives the opportunity for such a process.

Fig. 6.3 Copper indium diselenide fuel cycle

The waste cadmium and tellurium from the manufacturing plant can be economically recycled and emissions reduced to very small levels even for 100 MW$_p$ per annum production rates. Cadmium dust which might be present in the manufacturing plant could present a chronic hazard to personnel. However, the monitoring and control of such dust is well established in existing industries where cadmium is used, and no additional problems can be foreseen for PV manufacture.

The quantity of cadmium in the CdTe modules increases the potential hazards during operation and disposal compared to those with CIS cells. In operation, the modules can leak cadmium into the environment only if they are broken or subject to high temperatures. Broken modules pose little hazard as the area of cadmium exposed by even a break across the entire module would be less than 3 mm², and broken modules would be replaced quickly in routine maintenance. A fire in a building clad with CdTe modules could pose some hazard, although for small areas the hazard would be significant only for those so close to the fire as to suffer far greater hazards from smoke and flames. A large building having a large area of PV cladding could emit cadmium vapour—turning to cadmium oxide dust—into the smoke plume. Moskowitz *et al.* (1990) have studied this issue and conclude that standard fire-safety procedures would protect the population from this cadmium hazard, just as they protect them from the carcinogens in the smoke plume.

Disposal of CdTe modules may need to be controlled more strictly than for CIS modules, and recycling of materials is more important both for environmental reasons and for the value of the cadmium and tellurium. The technological problems associated with this recycling appear to be readily solvable at low cost.

Table 6.2 Hazardous emissions from photovoltaics

Material	Production	Operation	Disposal
Silicon	Silica dust Silanes Diborane Phosphene Solvents		
Copper indium diselenide	Hydrogen selenide Cadmium oxide Cadmium dust Selenium Solvents	Cadmium Selenium (in a fire)	Cadmium Selenium (if not recycled)
Cadmium telluride	Cadmium oxide Cadmium dust Tellurium Solvents	Cadmium Tellurium (in a fire)	Cadmium Tellurium (if not recycled)

Note: there is an energy input to all technologies derived from the fuel mix of the nation producing the PV.

The hazardous emissions from the various different types of solar cells are summarised in Table 6.2.

It is interesting to note that the amount of cadmium contained in the CdTe modules needed to generate energy of, say, 1 GWh over their lifetime is about equal to the cadmium emitted from the smoke-stack of a typical coal-fired station whilst generating the same 1 GWh of electrical energy. Very little of the cadmium in the PV modules would be lost into the environment, so the PV plant is cleaner, even for cadmium emissions, than a coal-fired power station.

Summary

The methodology established in the previous sections has been used to discuss the environmental impacts of photovoltaic power generation. We have discussed the manufacturing technology for cells and modules which have been described in Chapter 3: silicon, and CIS and CdTe technologies.

6.5 CONCLUSION

Photovoltaic systems are almost entirely benign in operation, and potential environmental hazards occur at the production and disposal stages.

Silicon is a very stable material and its release into the environment poses no hazards. In the production of silicon cells the hazards are similar to those encountered in the microelectronics industry, and monitoring and control procedures are well established for even the largest production rates envisaged.

For PV modules based on CIS or CdTe cells, there are potential hazards from the use of hydrogen selenide and cadmium. It is clear from this review, however, that there are well established methods of monitoring, control and alleviation which can reduce these hazards to within international safety limits for all stages of production, operation to disposal.

All production processes require an input of energy, and this energy is derived from the standard fuel mix of the nation in which production takes place. The production of PV systems has associated with it, therefore, emissions of greenhouse and acidic gases. In the present state of the PV industry, with small-scale production and technologies some way from maturity, these emissions are much less than those from fossil fuels but exceed those from the operation and fuelling of large nuclear reactors. As the new thin-film PV technologies come into production later in this decade, and the scale of production increases, the energy input to PV systems will decrease considerably, with consequent reduction in carbon dioxide emissions, to levels below that of other electricity generating technologies.

Environmental costs arise from many factors, and a valid comparison of the external costs of different energy sources must include all of these factors. There has been a significant effort in Europe and the USA to derive the external costs of electricity generation (Hohmeyer and Ottinger, 1991) and there is a growing consensus on the methodologies used in these

Fig. 6.4 Environmental effects of different energy technologies

calculations and on the external costs associated with the various techno-logies. Baumann and Hill (1991) have briefly reviewed these methodologies and have produced a matrix which allows a qualitative comparison of the magnitudes of the various components of external costs for the various energy technologies. This matrix is reproduced as Fig. 6.4, and it is clear that PV is the most environmentally benign of all electricity generation technologies presently envisaged for large-scale use throughout the world.

SUMMARY OF THE CHAPTER

The methodology for environmental assessment of energy sources has been reviewed and used to examine the environmental impact of photovoltaic power generation.

The analysis leads to the conclusion that the production and disposal of large PV generators poses significantly lower environmental risks or dam-ages to the environment than the conventional power generation techno-logies. The hazards during their operation are practically negligible.

BIBLIOGRAPHY AND REFERENCES

BADAWI, M. H., HYLAND, M., CARTER, M. J. and HILL, R., *Proc. 10th EC Photovoltaics Solar Energy Conf.*, Kleuver, Dordrecht, 1991, pp. 883–886.
BAUMANN, A. and HILL, R. *Proc. 10th EC Photovoltaics Solar Energy Conf.*, Kleuwer, Dordrecht, 1991, pp. 834–837.
BERNOW, S., BIEWALD, B. and MARRON, D., *in:* HOHMEYER, O. and OTTINGER, R. L., *eds. External Environmental Costs of Electric Power*, Springer, Berlin, 1991, pp. 81–102.
EC, *Energy and the Environment*, Brussels, 1990.
FRITSCHE, U., *in:* HOHMEYER, O. and OTTINGER, R. L., *eds. External Environmental Costs of Electric Power*, Springer, Berlin, 1991, pp. 191–209.
HOHMEYER, O. and OTTINGER, R. L., *eds. External Environmental Costs of Electric Power*, Berlin, 1991.
HYNES, K. M., PEARSALL, N. M. and HILL, R., *Proc. 10th EC Photovoltaics Solar Energy Conf.*, Kleuwer, Dordrecht, 1991, pp. 461–464.
MOSKOWITZ, P. D., ZWEIBEL, K. and FTHENAKIS, V. M., *Health, Safety and Environmental Issues Relating to Cadmium Usage in PV Energy Systems*, SERI/ TR-211-3621, Golden, CO, 1990.
OECD, *Environmental Effects of Electricity Generation*, Paris, 1985.
OECD, *Environmental Impacts of Renewable Energy*, Paris, 1988.
OECD, *Economic Instruments for Environmental Protection*, Paris, 1989.
PALZ, W. and ZIBETTA, H., *Int. J. Solar Energy* **12**, 3/4, 1991.
USDoE, *Energy Systems Emissions and Material Requirements*, Washington, DC, 1989.
ZWEIBEL, K., *Harnessing Solar Power—the Photovoltaic Challenge*, Plenum Press, New York/London, 1990.

SELF-ASSESSMENT QUESTIONS

1 What is the difference between 'direct' or 'internal' costs and 'indirect' or 'external' costs?

2 What are the major components of external costs?

3 What are the linkages between the operation of a technology and its environmental cost?

4 Where do the hazards arise in the production of silicon cells?

5 What issues are raised by the production of thin-film cells?

6 Under what circumstances could PV modules be a hazard during their operational lifetime?

7 Why should modules of thin-film polycrystalline cells be recycled?

8 How does PV compare in its environmental impact with other technologies for generating electricity?

Answers

1 Direct (internal) costs are those which are paid by the company and ultimately by the purchaser. They include energy, materials and equipment used in production, labour costs, and costs of capital. Indirect (external) costs are those incurred by production but not paid by the producer. They are paid by the society, or a section of society. Such costs include acid rain, air pollution by transport, the climatic change from the greenhouse effect, etc.

2 Environmental costs affecting the well-being of people and other living creatures and our rural and urban environment, and social costs which affect our well-being by inequitable allocation of resources.

3 *Operation* produces *emissions*. *Emissions* disperse into the environment and cause damage or *Burdens*, e.g. SO_2/m^2. The damage (*Burdens*) cause an *impact* on the environment, e.g. crop yields fall, or forests die. These *impacts* incur costs—the *environmental costs*.

4 During production of silicon from silica—emissions of silica dust. During manufacture of cells—phosphine, diborane, etc. During cleaning and etching operations and disposal of cleaning and etching chemicals.

5 For a-Si, the use of silane, phosphine and diborane. For CIS and CdTe, the use of heavy metals, particularly Cd, and for CIS, the use of H_2Se.

6 PV modules can be an electrical hazard during maintenance in bright sunlight; they can fall on someone whilst being fixed to the array. CIS and CdTe modules could emit Cd during fire.

7 Heavy metals are expensive, scarce and toxic, and should not be put in land-fill sites. They should be recovered and re-used at the end of the module lifetime.

8 Given the small and controllable impact per unit energy output of the PV modules in production, and their almost hazard-free operation and disposal, PV is perhaps the electrical energy generation technology with the lowest environmental impact of all those in use today.

7

Advanced and Specialised Topics

AIM

The aim of this unit is to discuss three individual topics of more specialised nature: large PV systems, photovoltaics under concentrated sunlight and the hydrogen economy.

OBJECTIVES

After completing this unit the student should be able to:

1. Understand the specific features of large PV systems, and examine their economic and technical characteristics.
2. Explain the advantages and disadvantages of concentrator PV systems, and discuss their operation.
3. Analyse the structure of the hydrogen economy, and show its feasibility and future potential.

NOTATION AND UNITS

Symbol		SI Unit
FF	fill factor	
I	current extracted from solar cell	A
I_{sc}	short-circuit current	A
I_o	dark saturation current	A

Written by A. Sorokin (Large PV projects), P. Davies and J. C. Miñano (Photovoltaics under concentrated sunlight), and M. Specht (Hydrogen economy).

(Contd.)

Symbol		SI Unit
k	Boltzmann constant	J/K
n	refractive index	
P_{loss}	power loss in series resistance	W
q	electron charge	C
R_s	series resistance	Ω
T	temperature of cell	K
V_{oc}	open-circuit voltage	V
V_{DC}	DC bus-bar voltage	V
X	optical concentration ratio	
X_{max}	maximum possible value of X	
G	irradiance	W/m^2
η	efficiency of solar cell	
θ_c	collection angle	
θ_e	angle of tracking error	
θ_{sun}	half-angle subtended by the sun	

Values of physical constants

$$k = 1.38 \times 10^{-23} \text{ J/K}$$
$$q = 1.602 \times 10^{-19} \text{ C}$$
$$\theta_{sun} = 0.27°$$

7.1 LARGE PV PROJECTS

7.1.1 Introduction

Unlike conventional power technologies where large-scale production facilities are generally measured in megawatts, photovoltaics has not yet reached such a large scale of applications.

The world's largest PV plant constructed so far, the Carissa Plains PV power plant in California, USA (Fig. 7.1), reached 5 MW installed peak power capacity. The largest PV plant in Europe, Kobern-Kondorf in Germany, has 500 kW capacity.

At present, the threshold limit value between large- and medium/small-scale PV systems may be set around 25 kW of installed PV peak power. As we have discussed in Chapter 5, this apparently low value in comparison with the traditional power technologies is due not only to the higher cost but also to area considerations.

7.1.2 Applications of large PV projects

Large PV projects have been built to provide electricity in many different applications (Table 7.1). It should be borne in mind that many of these

Fig. 7.1 Carissa Plains PV power station (courtesy of P. D. Maycock, Energy Systems Inc.)

applications were demonstration projects, built to study the feasibility as well as the technical and economic potentials of this new technology.

The experience from these applications has demonstrated that, unlike small-scale applications, large PV projects are less cost-effective as stand-alone than in grid-connected operation. This is due to the fact that grid-connected configurations can utilise all the produced PV energy, thus reducing the unit cost of generated electricity. Stand-alone applications, on the other hand, are heavily conditioned by the actual power demand of the isolated user which often means that the excess power cannot be utilised.

Table 7.1 Applications of large PV projects

- Grid-connected power production
- Stand-alone electrification of villages or communities
- Switchable electrification allowing both isolated stand-alone and grid-connected operation
- Water desalination and treatment (stand-alone)
- Water pumping (stand-alone)
- Cold-storage for food conservation (stand-alone)

7.1.3 Economics of large PV projects

There is little doubt that, at present, large PV projects are generally not competitive with the conventional power sources. The principal reasons are as follows:

- Investment costs are much higher, leading to much higher unit cost of generated electricity (see Table 7.2).
- Large PV power plants present technical complexity similar to the conventional power plants, requiring skilled personnel for their management and maintenance. Unlike small-scale PV systems they are therefore inappropriate for application in less developed remote areas as they cannot provide the simple and reliable power source that is required in these locations.
- Conventional power facilities take advantage of the economy of scale, i.e. the larger the facility the lower the specific production cost (the cost of a unit of generated electricity). Large PV plants, on the other hand, do not present similar advantages since, under given climatic conditions, the required active surface of PV modules, support structures and other facilities is simply proportional to the energy output of the plant.

Table 7.2 Cost of electricity generation

Technology	Investment costs		Unit cost (US$/kWh)
Large thermal power plant	1.5	US$/W	0.06
Diesel power plant	0.5–1.0	US$/W	0.2–0.5
Large PV power plant	12–15	US$/W$_p$	0.8–1.0

For these reasons, the industrial applications of large PV projects have been, with a few exceptions, limited to industrialised countries where government or transnational programmes have been able to promote the development of PV technology specifically in large-scale applications (mostly grid-connected power plants).

7.1.4 Technical characteristics

Large PV plants are generally characterised by the following features:

- High-voltage DC bus (typically 200–600 V_{DC}).
- Large-scale PV plants are generally AC systems which employ one or more inverters for the conversion of the DC power generated by the PV array to standard AC power.

A. Variable frequency AC system without battery (stand alone)

B. Fixed frequency AC system with battery (stand alone)

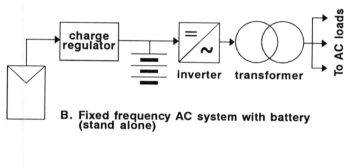

C. PV power plant (grid connected)

D. PV system using grid buffering

Fig. 7.2 Different types of large PV systems

The different types of large PV systems are shown schematically in Fig. 7.2. In common with other PV systems, they can be utility grid-connected or stand-alone. Stand-alone systems may, or may not, include a battery. The installation of battery storage in grid-connected systems gives little advantage since the grid may be used as a much more effective, unlimited and cheaper energy storage.

7.1.4.1 Stand-alone PV systems

Variable-frequency AC systems without battery are used where the application requires exclusively drive/shaft-power and an external energy storage is available. A typical application is for water pumping into a reservoir. These systems are characterized by a special variable-frequency inverter powering an AC motor. The drive-motor speed and, consequently, the power demand of the user is adapted instantaneously to the available sun power and makes the maximum use of the available PV power. The main advantages of such systems are high efficiency, rugged design and the possibility to feed any off-the-shelf standard AC motor which can operate at a variable speed (typically pumps, fans, etc.).

AC systems with battery represent the standard configuration of a PV power plant supplying an isolated (island) grid and may serve to electrify a community in a remote area. In this case one or (preferably) more fixed-frequency inverters feed continuously (24 h per day) standard AC power into a distribution grid.

Hybrid systems: The cost-effectiveness of a stand-alone PV power plant can often be improved by taking advantage of a hybrid (PV/wind or PV/diesel) design. During prolonged low-irradiation (no-sun) periods, the diesel gen-set (or other power source) then allows either to charge the main battery of the system, or to supply directly the AC users.

7.1.4.2 Utility grid-connected systems

PV power plants feed PV power directly and instantaneously into a utility grid by means of one or more inverters and transformers. In most larger systems, line-commutated inverters are used for such purpose, equipped with a maximum-power-point tracker.

PV systems using grid buffering use the grid as energy storage (instead of a battery) to integrate the power demand of the load. During the day, excess PV power is fed into the grid while during the night the power required by the load is reclaimed from the grid. Obviously the user of such system may minimize the energy exchange with the grid (and relevant costs) by adapting consumptions to the available sunpower.

7.1.5 Subsystems

The structure of PV systems and their subsystems were discussed in detail in Chapter 4. Here we focus attention on various features that are specific to large PV systems.

7.1.5.1 PV generator

Selection of the DC bus voltage In general terms one can say that the optimum performance favours high voltage giving higher conversion efficiencies, and requiring smaller conduit sections and smaller (and less expensive) equipment.

However, the DC bus voltage cannot be increased in an unlimited way for the following reasons:

- *PV modules* are designed for low-voltage applications and do not present sufficient insulation distances between live circuit parts (solar cells) and the grounded metal frame as required for high-voltage applications.
- *Safety regulations* define voltage thresholds to distinguish between different voltage classes (low/medium/high) and relevant requirements to be satisfied by equipment in these voltage ranges.
- Present *inverter technology* accepts only limited DC voltage levels depending on type of power-switching device adopted. Presently only SCR thyristors allow DC voltage levels exceeding 1000 V_{DC}.
- Standard *switching devices* for DC voltages above 500 V_{DC} are difficult to find since commercial equipment is generally designed for AC application and not to assure interruption of the opening arc occurring in high-voltage DC applications.

7.1.5.2 PV array lay-out

As we have seen in Chapter 4, the power rating of the PV array and the DC bus voltage specification define directly the electrical series/parallel configuration of the PV array.

In some instances, it may be advantageous to subdivide a large PV generator into electrically separate subarrays, particularly in the following situations:

- Different types of PV modules are to be installed in the same plant.
- Different array sections have to be installed at different orientations.
- The application requires the possibility to partialise (section) the PV generator

Once the electrical array configuration is defined, the physical layout of the PV array in terms of array sections (corresponding to electrical subarrays) and rows (containing single or multiple strings) may be determined.

7.1.5.3 Utility grid interface of PV power plants

Large-scale PV power plants are generally connected to a medium-voltage utility grid section of typically 10–20 kV_{AC}. The components of the grid interface which are generally included for this purpose are summarised in Table 7.3.

Table 7.3 Grid-interface components

- Step-up transformer(s)
- Power factor correction (capacitors) if required (e.g. for SCR inverter)
- Filtering for power quality improvement (harmonics)
- Grid interface protections which monitor:
  - maximum current limit
  - voltage range (max/min thresholds)
  - frequency range (max/min thresholds)
- Grid interface instrumentation, including at least:
  - active energy counters (separate for in/outgoing)
  - reactive energy counters (separate for in/outgoing)

7.1.5.4 System Control

In the early days of automatic control, all control functions of an installation were usually concentrated in one powerful (and expensive) central computer. Such a solution was believed to be cheaper and simpler since powerful software was expected to do all the work.

Frequently, an entire installation became dependent on the correct operation of the process controller, a very sensitive component of low reliability. As a result, nobody was able to operate the plant manually without the computer. And even worse, if the computer failed, there was no way to find out what had happened.

Modern control strategies therefore tend to adopt distributed controls: all control functions are split up into single and independent functional blocks or loops which are implemented by front-end control devices installed inside or nearby the addressed equipment. The coordination between different units is assured via shared process parameters (for example, battery voltage or PV bus voltage). The functions of the central automatic process controller (when included) should be limited to the minimum, i.e., to supervise and coordinate operation of the overall plant.

The control system of a large PV plant should include the following *four levels of control*:

1. *Process instrumentation* measures the status and values of process parameters and transmits this information to a higher-level controller (low-level control device, process controller or data acquisition system, or an indicator for the human operator).
2. *Front-end (low-level) control devices* perform specific control functions (loops) not requiring high-level coordination with other plant components (inverter internal controls, power factor correction, etc.).

3. The high-level *'intelligent' process supervisor* coordinates and supervises the operation of the entire plant, and acts as an interface with the human operator.
4. The *human operator* must be able to obtain easily all required information to provide the highest-level supervision of the plant.

Data acquisition (logger) functions may also be executed by the automatic plant supervisor, since the hardware is usually capable of performing this role. However, a separate unit specifically tailored for this purpose is usually more appropriate for the maximum reliability of data monitoring.

7.1.6 Safety aspects of high-voltage PV systems

As a result of the high DC bus voltage, large PV systems require specific safety measures which include personnel protection and grounding of all metal parts, including PV module frames. The safety measures must bear in mind the principal consideration that *the PV generator cannot be switched off effectively during daylight.*

Table 7.4 PV plant protection

- *DC shorts*: blocking diodes installed on all parallel strings
- *AC shorts* to be counteracted up-stream (inverter side) and down-stream (grid interface)
- *Grid interface protection*:
    - PV plant from the grid
    - Grid from the PV plant
- *Monitoring of insulation failures* which are relatively frequent in PV modules
- *Lightning and surge protection*:
    - Effective grounding of all metal structures
    - Galvanic separation of circuits in different areas of the plant
    - Overvoltage (surge) protection
    - Grounded Faraday-cage shielding of sensitive control electronics
- *Fencing* to avoid theft and vandalism

Table 7.4 gives a brief outline of the main protection measures that PV power stations should be provided with. Some of these measures have already been discussed in other parts of this book but protection with respect to lightning strikes should be particularly noted. Experience shows that lightning damage is usually suffered by the sensitive power conditioning and control electronics, rather than by the PV modules themselves. The lightning is most likely to strike the support structures of the PV array which are generally made of galvanized steel and well grounded. The utility grid also suffers frequent lightning strikes and conveys part of the resulting surges through the grid interface into the PV power plant.

Summary

The special features of large PV systems have been reviewed, and their applications and economics contrasted with smaller applications and conventional power sources. The technical characteristics were discussed with focus on the subsystems. It was shown that large PV generators are usually characterised by high DC bus voltage, leading to specific safety and protection measures. Other components of large PV systems, the grid interface and system control, have also been examined.

7.2 PHOTOVOLTAICS UNDER CONCENTRATED SUNLIGHT

7.2.1 Introduction

Concentration of sunlight is easily demonstrated by holding up a magnifying glass to the sun. The bright focus of light formed is intense enough to set fire to paper or wood. Place a photovoltaic cell at this focus, and you have a concentration photovoltaic system. Equally well, the concentrated energy could be used to drive a heat engine, as is the case with solar thermal systems.

In the case of photovoltaics, why should we be interested in using a concentrator? The main reason is to reduce the cost of the cell by making it smaller, whilst we still collect light from a large area (Fig. 7.3). For this to make sense, the concentrator has to be cheaper than photovoltaic cells of the same area. Another reason is that, once we are producing electricity from small cells, it becomes attractive to use more sophisticated, higher-efficiency types of cell which would be too expensive to use without concentrators. Such is the case with gallium arsenide cells, for example. Some of the most exciting developments currently taking place in solar-cell research may find their first use in concentration systems.

Cells designed to work in concentration systems produce a higher efficiency than is possible with unconcentrated sunlight. In Fig. 7.4 we see how

Fig. 7.3 (a) A normal solar cell in unconcentrated sunlight. The concentration cell (b) is smaller and cheaper but the concentrator has to be paid for

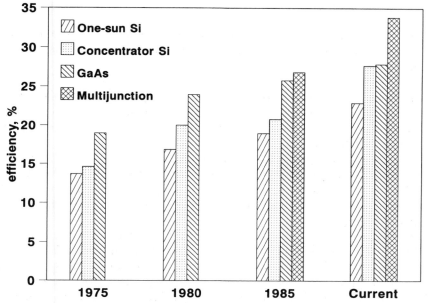

Fig. 7.4 Record efficiencies with and without concentration: silicon, and concentration GaAs and multijunction cells (after E. C. Boes and A. Luque, in T. B. Johannson, H. Kelly, A. K. N. Reddy and R. H. William, *Renewable Energy—Sources for Fuels and Electricity*, Earchscan: Island Press, London/Washington DC., 1993)

the record efficiencies for cells with and without concentration have risen since 1978. The highest efficiencies over 30%, achieved since 1988, are a result of using two types of cell stacked on top of each other, thus making good use of a broader part of the solar spectrum (see Chapter 3). So far 34% efficiency has been obtained in this way, using 100 times the concentration of sunlight. With three cell types and concentration, researchers hope to raise the figure to 40% in the next few years.

The efficiencies for cell and concentrator together are lower, due to loss of light in the concentrator and to cell heating, which both decrease efficiency. The record for prototype photovoltaic modules is therefore only about 23%.

Certain drawbacks have so far prevented concentration systems from making a significant penetration into the world photovoltaic market. In particular:

- Concentrators only use the direct beam light but not the diffuse light scattered by clouds and the atmosphere. This means that they are only useful where there is a large amount of direct sunlight. Even for a cloudless sky, the diffuse light can constitute 15–30% of the total solar radiation available.
- Concentrators always have to be pointed towards the sun. The mechanical system producing the required motion is called a *sun tracker*. The

additional complexity of the tracker has so far disfavoured use in remote locations because of reduced reliability and increased maintenance needs.

These two drawbacks do not apply to certain types of *static concentrators*, but these are not capable of concentrating sunlight more than about 10 times, whereas tracking concentrators can achieve much higher concentrations as we shall see.

Given these considerations, concentration systems are most likely to have significant advantages over conventional systems for central power generation (in sunny locations) for national or local grids (sections 5.7 and 7.1). Here, greater complexity can be justified and the constraints on cost and efficiency are most stringent.

The two largest concentration photovoltaic systems to have been built to date use Fresnel lenses and each have a peak output of about 300 kW. The first was completed in Saudi Arabia in 1981 and has an efficiency of about 9%. The second was installed in Texas, USA, in 1989, on the roof of a garage. Its efficiency is about 13%. These efficiencies are calculated from the direct current output of the system, and relative to the direct component of sunlight.

A number of other systems of output power greater than 10 kW have also been constructed. The power output of these installations is small compared to that of solar–thermal stations, the largest of which produces up to 100 MW. But perhaps this needs to be set against the fact that photovoltaics has only been a subject of serious research since the 1950s, whereas the heat engines used in solar–thermal installations have been under development for over two centuries.

7.2.2 Concentration and efficiency

Before studying optical systems which can be used to concentrate sunlight, we shall look at how solar cells behave under concentration.

In Chapter 3 we have seen that the short-circuit current I_{sc} is proportional to the irradiance of the cell. Under concentration, therefore,

$$I_{sc}(X) = X\,I_{sc}(1) \qquad (7.1)$$

where we denoted explicitly the dependence on the *optical concentration ratio*, X. The open-circuit voltage, on the other hand, is given by

$$V_{oc}(X) = (kT/q) \ln\left(\frac{I_{sc}(X)}{I_o} + 1\right) \approx (kT/q) \ln\left(X\,\frac{I_{sc}}{I_o}\right)$$

$$= V_{oc}(1) + (kT/q) \ln X \qquad (7.2)$$

where I_o is the diode dark saturation current, q is the magnitude of the electron charge, T is the temperature and k is the Boltzmann constant. Therefore, noting that the irradiance $G(X)$ increases in proportion to X, the efficiency as a function of X is:

$$\eta(X) = \frac{V_{oc}(X)I_{sc}(X)FF}{G(X)} = \eta(1)\left(1 + \frac{(kT/q)\ln X}{V_{oc}(1)}\right). \tag{7.3}$$

The relative increase of efficiency of the ideal solar cell is therefore proportional to the natural logarithm of the concentration ratio. For example, concentrating light tenfold $(X = 10)$ gives the voltage increase $(kT/q)\ln X$ equal to $0.026 \times 2.3 = 0.06$ V at 25°C. Since the open-circuit voltage of a silicon cell is typically 0.6 V without concentration, this amounts to about a 10% increase.

The efficiency of practical devices, however, cannot increase indefinitely. As we have seen in Chapter 3, a solar cell has a finite series resistance R_s. Power dissipated as heat in the series resistance is:

$$P_{loss} = I^2 R_s. \tag{7.4}$$

The current I flowing from the cell is proportional to the concentration X. On the basis of equation (7.1), the power wasted is:

$$P_{loss} \cong X^2 I_{sc}(1)^2 R_s \tag{7.5}$$

which grows rapidly as X is increased. It can be shown (Luque, 1989) that the optimum concentration is about:

$$X \cong (kT/q)/I_{sc}(1)R_s. \tag{7.6}$$

For example, suppose a cell produces a current of 50 mA at 1 sun (1 kW/m²) and has a series resistance of 0.01 Ω. With $kT/q = 0.026$ V (at 25°C), we can estimate the optimum concentration ratio X to be about 50, corresponding to an irradiance of 50 kW/m².

Although cells designed for concentration incorporate special features to reduce the series resistance, it is difficult to make a silicon cell with good efficiency at values of X much greater than about 200 (i.e. 200 kW/m²).

7.2.3 Maximum possible concentration

The maximum concentration ratio that can be achieved is affected by three principal factors:

- On a fundamental level, one must make allowance for the fact that the sun's rays that arrive at the Earth are not exactly parallel.

- If the cell is adjoined to a refractive medium, such as glass, this can make higher concentrations possible.
- In practical situations, the concentration ratio is also limited by tracking errors of the system.

These factors give the following limit on the concentration ratio:

$$X_{max} = n^2/\sin^2\theta_c \qquad (7.7)$$

where $\theta_c = \theta_{sun} + \theta_e$, θ_{sun} is the half-angle subtended by the sun's rays, θ_e is the tracking error (typically in the range 0.1° to 3°), and n is the refractive index of the adjoining medium (n is not usually greater than 1.6 for economic concentrators). The principle behind this formula is that sunlight cannot be concentrated beyond the point where it could heat an object to a temperature higher than that at the surface of the sun itself (Winston, 1991).

Since $\theta_{sun} = 0.27°$, the limit in the case of $n = 1$ and $\theta_e = 0$ (i.e. a solar cell surrounded by air in a system with no tracking errors) is $X_{max} = 45\,000$. The record concentration to have been achieved experimentally is 84 000, using sapphire for which $n = 1.76$ (Cooke et al., 1990). The theoretical limit for this refractive index is 140 000.

Concentrators which attempt to reach very high concentrations (> 200) usually have rotational symmetry, like those in Fig. 7.5. Concentrators of linear symmetry, such as those shown in Fig. 7.6, are often simpler to construct because they can be made by bending flat materials, but they achieve lower concentrations. The limit in their case is:

$$X_{max} = n/\sin\theta_c \qquad (7.8)$$

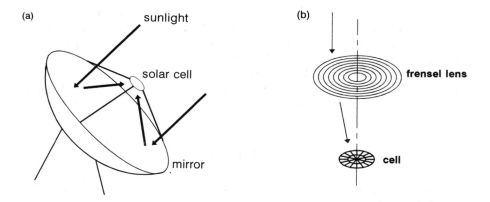

Fig. 7.5 Point-focus systems with rotational symmetry. (a) Dished parabolic mirror; (b) Fresnel lens

(a)

cells

mirror

(b)

lens

cells

Fig. 7.6 Line-focus systems with linear symmetry. (a) Parabolic mirror; (b) Curved Fresnel lens

which is only the square root of X_{max} as given by equation (7.7).

7.2.4 Design of practical concentrators

We shall now look at some real concentrators to see how they compare with the ideal maxima of equations (7.7) and (7.8). We often find that the concentrations obtained are much lower. However, this is not a great problem because the ideal limits are, to begin with, usually higher than the optimum of equation (7.6).

The concentration ratios we present, in terms of the collection angle θ_c, are those which could be obtained if there were no losses of light in the concentrator. In practice losses amount to at least 10%, due to absorption and Fresnel reflections, depending to a large extent on the quality of the materials used.

7.2.4.1 Focusing systems

Focusing systems use parabolic lenses or convex mirrors to focus parallel rays on a spot or line, as shown in Figs. 7.5 and 7.6. Because, however, the sun's rays are not absolutely parallel, the concentrated light does not come to a sharp focus but falls within a small area. Also, tracking inaccuracy can cause this illuminated area to be shifted off centre. The cell must be big enough to capture the illuminated area in all its possible positions, and this reduces the concentration ratio. However, a very large tracking error is always unacceptable for these systems because, even if they are designed to take it into account, the result is that just a small part of the cell is illuminated. Such gross non-uniformity of illumination decreases considerably the efficiency—just like a non-uniformly illuminated flat panel, as we saw in Chapter 4.

Let us now look at a practical example. For a dished parabolic mirror, like that of Fig. 7.5a, with the focal length of the mirror properly chosen, the maximum concentration obtainable is, for small collection angles (Welford and Winston, 1989).

$$X_{max} = 1/(4\,\theta_c^2) \qquad\qquad (7.9)$$

which is just a quarter of the maximum given by equation (7.7). Fig. 7.7a shows the variation of concentration with collection angle θ_c for the dished parabolic mirror, in comparison with the ideal limit; in Fig. 7.7b we show the situation for the line-focus system, where the concentration is just the square root of that given by equation (7.9).

The concentrations possible with mirrors are quite adequate for use with solar cells. In the case of the dished mirror (i.e. rotational symmetry), they tend to be higher than desired, although this is not an obstacle because the concentration can always be reduced by moving the cell closer to or further away from the mirror.

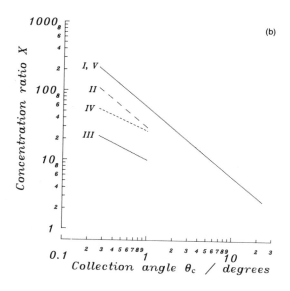

Fig. 7.7 Concentration ratio X as a function of the collection angle θ_c for concentrator with (a) rotational and (b) linear symmetry. I: Theoretical limit for $n = 1$ (equations (7.7) and (7.8)). II: Limit for parabolic mirrors. III: Limit for flat Fresnel lenses. IV: Limit for shaped Fresnel lenses (not shown in (a)). V: Limit for compound parabolic concentrators. VI: Example of a two-stage system using non-imaging secondary concentrator (Fig. 7.10a). VII: Example of a two-stage system using imaging secondary concentrator (Fig. 7.10b)

However, mirrors have some important drawbacks. Because the cell is mounted in the sunlight, it tends to be more difficult to cool it. The cell and cooling fins cast a shadow on the mirror, which decreases the power output of the system.

These factors have tended to favour Fresnel lenses over mirrors. The advantage of Fresnel lenses over solid lenses is that they are thinner and so use less material and are cheaper.

The equations for the maximum concentration are more complicated for lenses than mirrors. Here we shall only look at the results, which are summarized in Fig. 7.7a for point focus and Fig. 7.7b for line focus.

Point-focus Fresnel lenses (Fig. 7.5b) usually have a smooth, flat upper surface because this is easiest to fabricate and clean. Line-focus lenses (Fig. 7.6b) are easy to curve, which results in advantages in the concentration which may be obtained, as compared to flat line-focus lenses, although the concentration is always less than that obtained for point-focus systems.

The concentration obtained from lenses is lower than from mirrors, but still high enough to make practical systems. Indeed several such systems have been constructed. Fig. 7.8 shows a line-focus concentrator, like that in Fig. 7.6b, in which the lens and the side walls which connect the lens with the cells are made as a one-piece extrusion. This lowers the cost of the module, and provides a good protection for the cell against the environment.

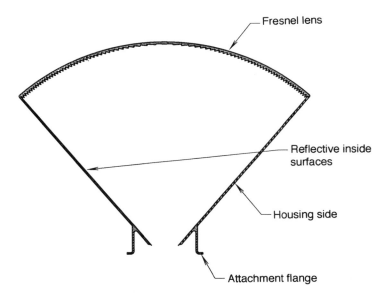

Fig. 7.8 Cross-section through a curved, linear Fresnel lens. The lens and housing walls are extruded in one piece (courtesy of SEA Corporation)

7.2.4.2 Non-imaging concentrators

The lenses and mirrors discussed above are similar in principle to those used in optical instruments such as telescopes, microscopes, photographic cameras, etc. All these instruments are used to produce an image. Similarly, focusing concentrators produce an image of the sun on the solar cell. The production of an image is, however, superfluous in the case of solar concentrators. All that matters is that the sun's rays should reach the cell. This approach leads to designs called *non-imaging concentrators*.

The most important type of non-imaging concentrator is called the compound parabolic concentrator (CPC), shown in Fig. 7.9a. It consists of two parabolic surfaces each having its focus at opposite extremes C and C′ of the cell. This means that rays entering the concentrator at an angle of $+\theta_c$ or $-\theta_c$ (measured to the axis of the concentrator) are reflected towards C and C′ respectively. By so designing the concentrator, we make sure that rays entering within a smaller angle reach the cell at some point between C and C′.

The concentrations reached by trough-shaped CPCs (i.e. those of linear symmetry as shown in Fig. 7.9b) are equal to the absolute limit given by equation (7.8). For CPCs of rotational symmetry (Fig. 7.9c), the concentrations are just slightly below the absolute limit (see Fig. 7.7a).

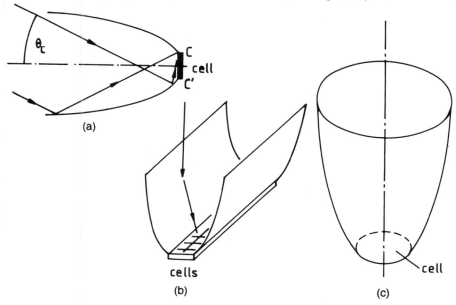

Fig. 7.9 (a) Cross-section through CPC showing how rays at an angle θ_c to the axis of symmetry are directed to one extreme C of the solar cell. Similarly, rays entering at $-\theta_c$ are directed towards C′. (b) Trough-shaped CPC of linear symmetry. (c) Funnel-shaped CPC of rotational symmetry

The possibility of using high acceptance angles and still achieving reasonable concentrations makes the trough CPC attractive for *static concentrators*. These do not use any tracking system, but make use of the fact that the elevation angle of the sun above a plane parallel with the equator is restricted to $\pm 23°$ about its mean value throughout the year. We can therefore accept this variation as a tracking inaccuracy. The result is that a concentration of about 2.5, (or greater if the cell is adjoined to a medium of refractive index higher than unity), can be obtained. Larger concentrations are possible if small adjustments are made manually to the orientation of the concentrator throughout the year. This might be attractive in locations where technological resources are limited.

A drawback of the CPC is that, for high concentration, they have to be very long in relation to the diameter of their entrance aperture. This makes them unwieldy and expensive for concentration ratios above about 50. Another disadvantage of the CPC is that it does not guarantee uniform illumination of the solar cell.

7.2.4.3 Two-stage concentrators

Almost all of the disadvantages of focusing and non-imaging concentrators are overcome by using concentrators having two stages, called a primary and secondary concentrator. The relation between concentration and acceptance angle can be close to the ideal limit, the problem of non-uniformity of illumination can be corrected, and the concentrator can be quite compact.

The only disadvantages are the extra complexity and cost, and the additional optical losses which can occur in the second stage.

A focusing optical element, such as a flat Fresnel lens, is generally used for the primary concentrator. The secondary can either be a non-imaging mirror (see Fig. 7.10a), similar in principle to the CPC of Fig. 7.9c, or an imaging lens (see Fig. 7.10b). The non-imaging type gives very good tolerance to tracking errors, whereas the imaging secondary is particularly attractive in allowing a very uniform illumination of the cell.

Because so many different designs of two-stage concentrator are possible, it is not easy to summarise their performance. However, we show in Fig. 7.7a the concentrations and collection angles for the examples of Figs. 7.10a and Fig. 7.10b. Compared to the CPCs achieving equivalent concentrations, their length is reduced 3 to 9 times.

7.2.5 Tracking systems

We have seen how (apart from the static concentrators) tracking systems are needed in concentration systems, and the more accurate the tracking, the greater the concentration which can be obtained. In Chapter 2, the motion of

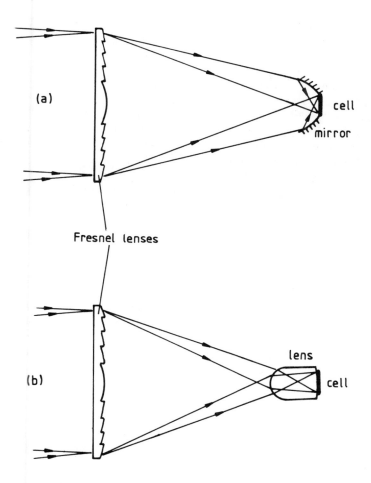

Fig. 7.10 Two-stage concentrators in which the primary stage is a Fresnel lens. (a) Non-imaging secondary stage similar to CPC. (b) Imaging secondary stage consisting of a single-surface lens

the earth around the sun and its effect on sunlight have been described. What the tracker has to do is move the concentrator modules in such a way as to cancel out this motion, keeping their orientation relative to the sun fixed.

In the polar-axis system, for instance, the main axis of rotation is parallel to the earth's polar axis. The rotation about this axis is at the same speed as the earth's rotation, but in the opposite direction. Then, throughout the course of the year, it is necessary for the panels to be tilted relative to the polar axis because of the Earth's declination.

Fig. 7.11 Polar axis tracking system being used to pump water in Bangladesh. The heat of the sun is used to produce the tracking motion via a mechanical system (courtesy of Midway Labs, Inc.)

A practical example of a polar-axis system is shown in Fig. 7.11. An interesting feature of this design is that, unlike most trackers which are driven by electric motors, it is worked by fluids heated directly by the Sun. The heat causes the fluids to expand and move pistons.

A different type of tracking mechanism is the azimuth-elevation type, an example of which we see in Fig. 7.12a. The azimuth axis is vertical. At sunrise the concentrator modules face east, then during the day rotation around the azimuth axis turns them round to face west at sunset. At the same time, the elevation axis (horizontal and mounted on top of the azimuth axis) tilts the module upward as the sun rises up till noon, then down again as dusk approaches. An interesting feature of this system is the way the modules are staggered, leaving spaces for the wind to pass through. This reduces the wind loading, and thus the strength required of the structure. Fig. 7.12b shows the module used in this machine.

Although the motions of the azimuth-elevation tracker are more complicated and difficult to control than those of the polar-axis type, an advantage seems to be that the structure does not have to be so high (relative to the module size) to avoid the modules hitting the ground.

The above examples both use two axes of motion. One-axis systems are possible with non-imaging concentrators and some line-focus systems. In Fig. 7.13 we give an example, a venetian-blind type tracking system, used with the linear Fresnel lenses as of the kind shown in Fig. 7.8.

Summary

Concentration can be advantageous because of the reduction in the area of photovoltaic cells it achieves, but it introduces extra complexity in terms of the need to track the Sun. The output voltage and hence efficiency of cells tend to increase under concentration but, under very high concentrations, losses through the electrical resistance of the cell cause the efficiency to decline.

There is a theoretical limit to how many times light can be concentrated. The limit is lowered by tracking inaccuracy which is significant in all practical systems. Dished parabolic mirrors only achieve 1/4 of the theoretical limit and lenses achieve less. But lenses are still adequate for practical systems provided the tracking accuracy is good i.e. better than $\pm 1°$. Non-imaging concentrators such as the compound parabolic concentrator permit greater inaccuracy, but are unsuitable for large concentrations (> 50) because of their excessive size. Two-stage concentrators combine (at the expense of greater complexity) high concentration, tolerance to tracking inaccuracy, and small size.

Various tracking mechanisms have been built. The one-axis systems are simpler than the two-axis systems, but cannot be used to obtain such high concentrations.

7.3 ALTERNATIVE STORAGE: THE HYDROGEN ECONOMY

7.3.1 Introduction

As we have amply discussed throughout this book, the energy generated by a photovoltaic generator usually needs storage to compensate for the intermittent nature of solar radiation. The conventional energy storage by batteries was discussed in Chapter 4. In this section we describe how solar energy can be stored in an alternative chemical form, by hydrogen production from water. Indeed, as we shall see presently, this technology represents more than energy storage: it offers a new form of energy economy which complies with the current environmental requirements.

The advantages of using hydrogen as energy storage are summarised in Table 7.5. An important point to note is that the use of hydrogen does not generate harmful emissions to the environment. In particular, it does not produce carbon dioxide which contributes to the greenhouse effect. Pure carbon burns to CO_2, creating 0.34 kg of CO_2 per kWh when coal is used (kWh here refers to the thermal energy). Hydrocarbons burn to form CO_2 and water. The CO_2 emision is here about 0.27 kg in the case of oil and 0.2 kg for natural gas. In comparison, hydrogen combustion is CO_2 free.

2X4 LENS PARQUET (3)

5 BULKHEADS TO INCREASE
MODULE STIFFNESS

CELL ASSEMBLIES
CONNECTED IN SERIES

114" x 19.75" x 12"

MODULE OUTPUT: 150 WATTS

(b)

Fig. 7.12 A large azimuth-elevation tracking system designed to give 15kW power output. (a) (opposite) The modules are staggered to leave gaps for the wind to pass through which reduces the load on the structure. (b) Cross-section through the module used in this system which is similar in principle to that in Fig. 7.10b (both figures courtesy of Alpha Solarco)

In the course of the industrial development, there was a change from coal to other fuels with lower CO_2 emissions. It is envisaged that this trend will continue through the intensified use of natural gas with a high hydrogen content to the exclusive use of hydrogen which burns without CO_2 emissions. The H/C ratio of coal is about 1:1 whilst for the natural gas, it is 4:1. When hydrogen is used on a large scale, the H/C ratio of the energy carriers will be greater than 4:1.

Fig. 7.13 Venetian-blind type one-axis tracking system for line-focus Fresnel lenses of the kind shown in Figs. 7.6b and 7.8 (Courtesy of SEA Corporation)

Table 7.5 Advantages of hydrogen as energy carrier

- *Established technology.* Hydrogen generation by electrolysis is a well known process. Hydrogen is one of the oldest industrial gases and its handling is well understood.
- *Compatible with photovoltaic generation.* Water electrolysis requires a low-voltage DC source.
- *Ease of use and storage.* Hydrogen can be used directly for electricity, heat and power generation, and can be stored with minimal losses.
- *No global environmental risks.* Hydrogen is environmentally neutral. The reaction product is water (no emissions of CO_2 CO, SO_2, hydrocarbons, dust or ash). In addition to water, only traces of NO_x are formed in the combustion porocess.
- *Feasible generation and transport.* Hydrogen can be generated in solar farms in the sun belts of the earth, and transported to the consumer by pipelines or tankers.
- *Existing infrastructure can be used.* Natural gas can be successively replaced by hydrogen which will be fed into the existing natural gas pipeline system (in former times, town gas was used with a content of 50–60% hydrogen).

It is foreseen that the hydrogen economy would consist of the following components (Fig. 7.14):

- *Electricity generation* by photovoltaic plants or other renewable sources. It is important to choose an energy supply of the future that will not damage the environment. This can only be based on renewable energy sources.
- *Production of hydrogen* by electrolysis (section 7.3.2).

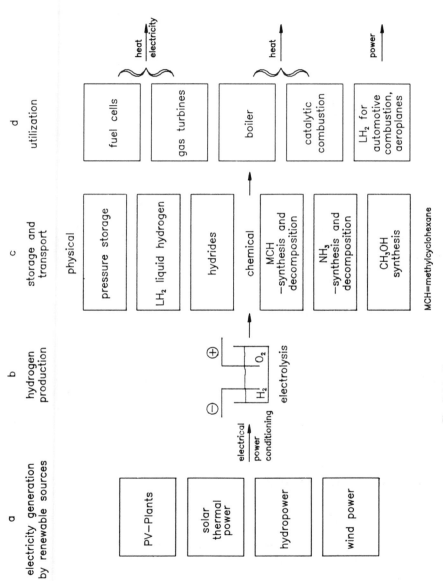

Fig. 7.14 Components of the hydrogen economy

● *Transport and storage of hydrogen.* The physical and chemical methods for local storage systems, automotive storage and intercontinental transport are discussed in section 7.3.3.
● *Utilisation of hydrogen*, for example, heating, electricity generation in fuel cells and gas turbines, or as transportation fuel (section 7.3.4).

alkaline electrolysis

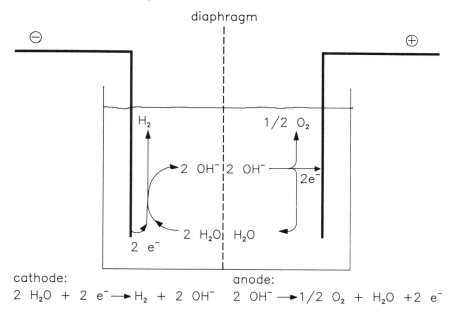

Fig. 7.15 Diagram of the conventional (or alkaline) water electrolysis

7.3.2 Generation of hydrogen

7.3.2.1 Electrolysis

Hydrogen can easily be generated by electrochemical splitting of water (*electrolysis*—see Figs. 7.15 and 7.16). When two electrodes are submerged in the electrolyte, the following two reactions take place:

At the cathode (negative electrode) electrons flow from the electrode to the electrolyte and *reduce* the hydrogen ions:

$$2H^+ + 2e^- \rightarrow H_2. \qquad (7.10)$$

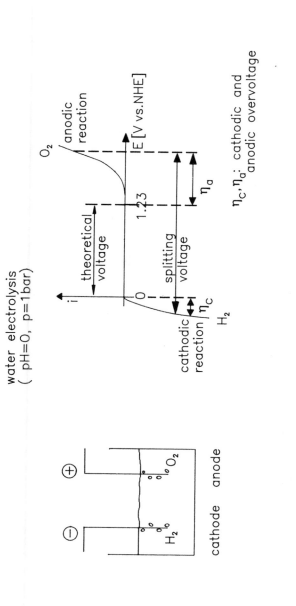

water electrolysis
(pH=0, p=1bar)

η_c, η_a: cathodic and anodic overvoltage

cathode: $2 \, H^+ + 2 \, e^- \longrightarrow H_2$ $\qquad E_0 \, (H_2/H^+) = 0.00 \, [V]$

anode : $\qquad H_2O \longrightarrow 1/2 \, O_2 + 2 \, H^+ + 2 \, e^-$ $\qquad E_0 \, (H_2O/O_2) = 1.23 \, [V]$

$\Sigma \quad ; \qquad H_2O \longrightarrow H_2 + 1/2 \, O_2$ $\qquad E_{H_2/O_2} = 1.23 \, [V]$

Fig. 7.16 The principles of water electrolysis. The voltage is referred to the normal hydrogen electrode, the zero point of the electrochemical scale

At the anode (positive electrode) electrons flow from the electrolyte to the electrode and promote the *oxidation* reaction

$$H_2O \rightarrow \tfrac{1}{2} O_2 + 2H^+ + 2e^-. \qquad (7.11)$$

The overall reaction is therefore,

$$H_2O \rightarrow H_2 + \tfrac{1}{2} O_2. \qquad (7.12)$$

The water splitting voltage, determined from the thermodynamics of reactions (7.10) and (7.11), is equal to 1.23 V. Due to electrochemical losses (overvoltages), the real voltage under current flow is usually greater than 1.6 V.

The energy efficiency of a water electrolyser can be determined from the following arguments. An electrolytic cell normally operates at current densities of a few hundred mA/cm². This is a compromise between economy and energy efficiency. The energy consumption for a conventional electrolyser is about 4.8 kWh/N m³ (kWh/norm cubic meter of hydrogen) but for an advanced alkaline technique this value is only 3.9 kWh/N m³. The theoretical value with 100% efficiency is 2.94 kWh/N m³ (referred to as the lower heating value of hydrogen) giving energy efficiency of an advanced electrolyser as (2.94/3.9) × 100 = 75%.

There is a number of ways in which electrolysis may be implemented in practice (Fig. 7.17). In the most advanced version of the conventional or *alkaline water electrolysis*, the electrolyte is 30% KOH where the ion transport takes place via OH⁻ and K⁺. A diaphragm is inserted between the two electrodes and prevents mixing of the hydrogen and oxygen gases. The electrodes are usually made from nickel, and the electrolysis proceeds at a temperature of 80°C.

In the electrolysis based on *solid polymer electrolysis*, the electrolyte is a proton-exchange membrane which conducts H⁺ ions. The operating temperature is usually between 80 and 120°C.

High-temperature electrolysis (at temperatures 800–900°C) is not available yet, but promises the greatest energy efficiency, converting more than 90% of the electric energy. The solid electrolyte, based on ZrO_2/Y_2O_3, conducts here the O^{2-} ions.

There are several requirements on the interconnection between the photovoltaic generator and the electrolyser, as well as on the electrolyser itself. The principal ones are as follows.

- Good dynamic response at intermittent load.
- High gas purity (H_2/O_2 mixtures form explosive gases).
- Optimisation of power conditioning between the PV generator and the electrolyser.

conventional electrolysis

electrolyte:
30 % KOH
(T ~ 80 °C)

solid polymer electrolysis

$$H_2O \longrightarrow 1/2\ O_2 + 2\ H^+ + 2\ e^-$$

$\Big\{$ electrolyte:
protone exchange membrane
(T : 80 − 120 °C)

$$2\ H^+ + 2\ e^- \longrightarrow H_2$$

high temperature steam electrolysis

$$H_2O + 2\ e^- \longrightarrow H_2 + O^{2-}$$

$\Big\{$ electrolyte:
ZrO_2/Y_2O_3
(T : 800 − 900 °C)

$$O^{2-} \longrightarrow 1/2\ O_2 + 2\ e^-$$

Fig. 7.17 The different types of electrolysis

The technology of continuous water electrolysis is well developed. However, discontinuous electrolysis powered by intermittent sources (photovoltaics or solar thermal generators) requires more research and development. During solar operation, the electrodes have to cope with a widely fluctuating load. This behaviour affects the activity of the electrocatalysts. Protection potentials have to be applied to the electrodes when there is no irradiation (at night). Futher design criteria for solar operation of an electrolyser are low specific energy consumption, high gas purity, and good current efficiency at partial load, as well as a good dynamic response at intermittent load in the entire current density range.

  Since the photovoltaic generators and electrolysers are modular, a variety of electric connections are possible. The simplest connection between the PV generator and electrolyser is by direct wiring with different string connections. Higher efficiencies are attainable by using a power-conditioning DC/DC converter with full MPP tracking. Up to 95% of the generator power can then be directed to the electrolyser.

7.3.2.2 Other hydrogen generation technologies

An alternative concept for hydrogen production is a direct conversion of light energy into chemical energy in a photo-electrochemical reaction (analogous to photosynthesis in biology). In contrast with the two-step photovoltaic electrolysis system, the photo-electrochemical reaction is a one-step process. An illuminated, hydrogen-evolving semiconductor photo-cathode can be combined with a photoanode to form a photo-electrolytic cell. In principle, this concept represents a simpler technological method, but, due to corrosion processes at the semiconductor electrodes, photo-electrochemical water-splitting systems are not available at present. Other direct solar-to-chemical energy conversion methods such as biological and thermal processes are still in the research stages and will not be available in the near future.

7.3.3 Transport and storage of hydrogen

The hydrogen storage technology can be divided into physical and chemical methods.

7.3.3.1 Physical methods

The simplest storage system for hydrogen is to fill pressure vessels and gas tanks with *gaseous hydrogen* (GH_2). This system plays an important role for the stationary storage of natural gas and hydrogen, and in pipeline systems. Because of the low energy density and high weight of the pressure vessel as well as the necessary safety precautions, this system is not suitable for automotive storage. For intercontinental hydrogen transport from the sun belts to the consumer, pipelines can be used in the same way as with the oil and natural gas transport today. This large transport system is also a storage medium which is able to buffer between the intermittent hydrogen generation and energy consumption.

The energy density of hydrogen can be dramatically increased by liquefaction. The temperature of *liquid hydrogen* (LH_2) is about $-253°C$ which requires the use of well insulated tanks. In comparison with the density of GH_2 at 1 bar (90 g/m^3), the density of LH_2 is 71 kg/m^3. Liquefaction of hydrogen, however, consumes large amounts of energy. One third of the inherent energy of the gaseous hydrogen is used for liquefaction. LH_2 therefore retains only about 55% of the electric energy used to produce it.

In the hydrogen economy, LH_2 will be used for intercontinental shipping in tankers, for hydrogen-powered airplanes, and for automotive storage.

Certain metals, in the form of *hydrides*, can store hydrogen embedded in their molecular lattice. These metal hydrides are of interest for hydrogen-powered cars (automotive combustion or electricity generation in fuel cells). The advantage of this system is a high charge and discharge efficiency, but the disadvantage is the high weight of the metal hydride storage.

7.3.3.2 Chemical storage methods

In addition to liquefaction, chemical storage is of interest for the intercontinental tanker transport of hydrogen. In many applications, the handling of liquid fuels at ambient temperature is easier than the low-temperature storage (see Fig. 7.14).

When combined with hydrogen, chemicals like toluene form new compounds. The reaction product is methylcyclohexane (MCH) in the case of toluene:

$$C_7H_8 \text{ (toluene)} + 3H_2 \rightarrow C_7H_{14} \text{ (MCH)}. \qquad (7.13)$$

MCH is liquid at ambient pressure and temperature and can be stored and transported in existing oil tanks and ships. After transport, the hydrogen carrier has to be dehydrogenated and the hydrogen is ready for use. Toluene has to be shipped back again for hydrogenation in a two-way transport. Another possible chemical storage of hydrogen is by ammonia production:

$$N_2 + 3H_2 \rightarrow 2NH_3. \qquad (7.14)$$

Liquid ammonia is contained unpressurised at $-33°C$, or at the normal temperature at a pressure of 10 bar. The low-temperature transport here does not represent the same problem as for liquid hydrogen. The overall efficiencies for hydrogen storage in the form of MCH (59%) and NH_3 (61%) are slightly higher than the LH_2 storage (55%).

The use of solar hydrogen as a chemical raw material for the production of synthetic hydrocarbons (synfuels with a high H/C ratio, e.g. methanol) is one of the most important topics in solar hydrogen storage research. One molecule of methanol contains four hydrogen atoms. Compared to other hydrogen energy carriers like methylcyclohexane or ammonia, methanol is easy to transport and handle, and is ready for use without any chemical decomposition processes. The methanol technology as an energy carrier is fully developed, and storage of liquid methanol is possible at the ambient temperature. Methanol offers a 1.8 times higher volume energy density than liquid hydrogen. An additional economic advantage would be the possibility of using the existing infrastructure of gas stations and transport systems for automotive combustion.

The combination of fossil and solar sources is feasible, for example, by combining synthesis gas with solar hydrogen to produce methanol:

$$(H_2 + CO)_{fossil} + H_2 \rightarrow CH_3OH. \tag{7.15}$$

The synthesis of methanol can be achieved by a catalytic chemical process between gaseous CO_2 and H_2:

$$CO_2 + 3H_2 \rightarrow CH_3OH + H_2O \tag{7.16}$$

or, in the future, by a direct electrochemical reduction in CO_2-containing electrolytes. The recycling of CO_2 will help to prevent the rise in the atmospheric concentration of this gas. In the race to develop a cleaner-burning fuel from renewable sources, methanol is a practical option for solar hydrogen storage. Calculated values of the overall energy efficiency of the synthesis, however, are slightly lower than in LH_2 generation.

7.3.4 Hydrogen utilisation technologies

Hydrogen has a wide range of uses for heat, power and electric generation (see Fig. 7.14). Some examples are listed below.

Fuel cells. In the hydrogen economy, fuel cells will play a major role. The energy efficiency of electricity generation directly from chemical energy is not limited by the Carnot process, and values in excess of 60% can be realised. The principle of a fuel cell is the reversal of electrolysis. Hydrogen is oxidised at the anode (which is here the negative electrode) and oxygen is reduced at the cathode (positive electrode). Figure 7.18 compares the current-voltage behaviour of a fuel cell with an electrolytic cell:

$$\text{anode:} \qquad H_2 \rightarrow 2H^+ + 2e^-$$

$$\text{cathode:} \qquad \tfrac{1}{2}O_2 + 2H^+ + 2e^- \rightarrow H_2O$$

$$\text{overall reaction:} \quad H_2 + \tfrac{1}{2}O_2 \rightarrow H_2O$$

The reaction product is water without any pollutants.

Catalytic burners. In comparison with flame combustion, catalytic combustion of hydrogen exhibits an inherent advantage of high reaction efficiencies combined with very low NO_x emissions. During flame-free combustion, hydrogen reacts with air when in contact with solid catalysts at relatively low temperatures. Catalytic heaters operate in a range from the ambient temperature to several hundred degrees Celsius.

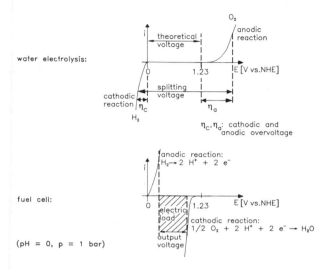

Fig. 7.18 Operation of the fuel cell

Hydrogen for steam generation. Steam generators can be used for peak load in conventional electric power plants and for industrial processes. The power can be increased from 0 to 100 MW in 1 s by a hydrogen/oxygen steam generator based on rocket technology.

Hydrogen in aviation. The advantage of using hydrogen in aviation is the weight of LH_2. With the same energy content, LH_2 weighs only one third compared to conventional aircraft fuels. The take-off weight is reduced by about 30% for the same payload and range.

Hydrogen as raw material. Hydrogen is also an important raw material in industrial synthesis processes (for example, ammonia synthesis for fertilisers, hydrogenation in petrochemical industry, methanol synthesis, reduction processes in metallurgy, etc.)

Summary

Hydrogen economy combined with photovoltaic generation offer the possibility of energy technology without carbon dioxide or other environmentally damaging emissions. The hydrogen generation by electrolysis is a well known process. The storage and transport are well understood and can build on the existing infrastructure of pipelines and tankers. These aspects, together with hydrogen utilisation technologies were discussed in detail, highlighting future potential of the technology.

BIBLIOGRAPHY AND REFERENCES

COOKE, D., GLECKMAN, P., KREBS, H., O'GALLAGHER, J., SAGIE, D. and
WINSTON, R., Sunlight brighter than the sun, *Nature* **346**, 1990: 802.
LUQUE, A., *Solar Cells and Optics for Photovoltaic Concentration*, Adam Hilger,
Bristol, 1989, p. 111.
WELFORD, W. T. and WINSTON, R., *High Collection Nonimaging Optics*,
Academic, New York, 1989, p. 50.
WINSTON, R., Non-imaging optics, *Scientific American* **264** (3) 1991: 52–57.

SELF-ASSESSMENT QUESTIONS

Large PV projects

True or false?

1 Large PV projects are measured on the scale of:
   (a) hundreds of megawatts
   (b) hundreds of kilowatts.

2 Large PV projects are generally more cost effective in:
   (a) grid-connected applications
   (b) stand-alone configurations.

3 Under favourable climatic conditions the unit cost of electricity generated by
   a large-scale renewable energy source is only slightly higher than that from a
   conventional power plant. This statement generally applies to:
   (a) wind-power
   (b) PV solar power.

4 Under favourable circumstances PV systems may be more cost-effective
   and compete with conventional power sources. This statement generally
   applies to:
   (a) small-scale stand-alone applications
   (b) large-scale grid-connected PV power plants.

5 For PV applications in the third world the following overall energy system
   conception should be preferred:
   (a) concentrated large-scale PV power plant feeding users through a conven-
        tional distribution grid
   (b) many diffused small-scale PV systems feeding relevant users locally.

6 According to the principle of 'economy of scale' a large production facility
   presents lower unit production costs (cost/unit) than a small production facility
   of otherwise same characteristics. Generally this principle applies also to:
   (a) Conventional large scale (bulk) power plants
   (b) Diesel power plants
   (c) PV module manufacturing
   (d) PV systems (applications).

7 In grid-connected PV power plants the installation of a battery yields significant advantages since it allows accumulation of excess energy during the day and supply to users (of the grid) also during the night.

8 In large-scale PV systems a high DC bus voltage generally yields higher system efficiency.

9 Considering the high reliability standards reached by modern computers, the best strategy for the automation of a PV plant is to provide for this purpose:
(a) A centralised and powerful process control system
(b) A distributed control system consisting of independent functional blocks and control loops.

Photovoltaics under concentrated sunlight

1 A choice is to be made between two photovoltaic systems:
(i) conventional panels costing $500 per m², and
(ii) a concentration system using lenses costing $200 per m² of area from which sunlight is collected, together with cells, whose area is only 1/100th of that area, costing $2000 per m² of cell area.
Each system produces the same average power output for a given collection area. On the basis of these figures, which is the cheapest? What other consideration might affect the choice in practice?

2 A perfectly designed lens could theoretically concentrate sunlight to an infinite extent—is this true or false?

3 What merits and drawbacks do line-focus concentrators have as compared to point-focus ones?

4 What do you understand by an *ideal concentrator*?

5 Why is a sun tracker more important for concentration systems than for ordinary photovoltaic panels?

Hydrogen economy

1 What are the main advantages of hydrogen as a solar energy carrier?

2 What are CO_2 emissions of different fossil energy carriers?

3 What are the major components of a solar hydrogen economy?

4 What are the reactions at the electrodes during water splitting in a conventional electrolysis cell?

5 What is the theoretical and practical splitting voltage?

6 How does one calculate the energy efficiency of a water electrolyser?

7 What are the main requirements on the electrolyser, and on the interconnection between a PV generator and an electrolyser?

Answers

Large PV projects

1a, False; 1b, True; 2a, True; 2b, False; 3a, True; 3b, False; 4a, True; 4b, False; 5a, False; 5b, True; 6a, True; 6b, True, 6c, True, 6d, False; 7, False; 8, True; 9a, False; 9b, True.

Photovoltaics under concentrated sunlight

1 The problem is to work out the cost of the second system per m^2 of collection area and compare it with the conventional system. For every m^2 of collection area we have in (ii) 0.01 m^2 of cells, costing \$20, which added to the cost of the lenses gives \$220/m^2 in total. This is cheaper than the conventional system. However, the sun-tracker would add to the cost of the second system and might affect the reliability.

2 False: the concentration is limited by thermodynamic considerations, however good the lens is.

3 Line-focus concentrators are generally easier to construct since they can more often make use of flat sheets of material bent into shape. Also, they may be used with tracking about one instead of two axes. However, the concentration ratios they can produce are much lower.

4 Ideal concentrators produce concentration ratios in relation to the angle within which they collect light, as high as the theoretical maxima of equation (7.7) (or equation (7.8) for concentrators of linear symmetry). No energy is lost in them, i.e. all light entering within the collection angle is transmitted to the solar cell.

5 Sun-trackers are used to keep solar concentrators pointed towards the sun as it appears to move through the sky. Unlike ordinary photovoltaic panels, concentration systems only accept light within a certain range of angles and so must be orientated correctly with respect to the sun.

Hydrogen economy

1 See Table 7.1.

2 Burning of coal, 0.34 kg CO_2 per kWh_{th}, 0.27 for oil, and 0.2 for natural gas.

3 Electricity generation by a renewable source, hydrogen production by electrolysis, transport and storage of hydrogen, followed by utilisation.

4 $2H^+ + 2e^- \rightarrow H_2$ at the cathode

$H_2O \rightarrow \frac{1}{2}O_2 + 2H^+ + 2e^-$ at the anode

The overall reaction is $H_2O \rightarrow H_2 + \frac{1}{2}O_2$

5 Theoretical splitting voltage is 1.23 V. In practice, voltage greater than 1.6 V is required.

6 Energy efficiency

$$= \frac{\text{Lower heating value of hydrogen (} = 2.94 \text{ kWh)}}{\text{Energy consumption in kWh per norm } m^3 \text{ of hydrogen}}.$$

7 Good dynamic response at intermittent load, high gas purity, and MPP tracker for best operation.

Index

Index compiled by Paul Nash

6084